TAKING LEAVE
OF DARWIN

TAKING LEAVE
OF DARWIN

A LONGTIME AGNOSTIC
DISCOVERS THE CASE FOR DESIGN

NEIL THOMAS

SEATTLE   DISCOVERY INSTITUTE PRESS   2021

Description

University professor Neil Thomas was a committed Darwinist and agnostic—until an investigation of evolutionary theory led him to a startling conclusion: "I had been conned!" As he studied the work of Darwin's defenders, he found himself encountering tactics eerily similar to the methods of political brainwashing he had studied as a scholar. Thomas felt impelled to write a book as a sort of warning call to humanity: "Beware! You have been fooled!" The result is *Taking Leave of Darwin*, a wide-ranging history of the evolution debate. Thomas uncovers many formidable Darwin opponents that most people know nothing about, ably distills crucial objections raised early and late against Darwinism, and shows that those objections have been explained away but never effectively answered. Thomas's deeply personal conclusion? Intelligent design is not only possible but, indeed, is presently the most reasonable explanation for the origin of life's great diversity of forms.

Library Cataloging Data

Taking Leave of Darwin: A Longtime Agnostic Discovers the Case for Design by Neil Thomas

166 pages, 6 x 9 x 0.4 & 0.5 lb, 229 x 152 x 9 mm & 0.235

Library of Congress Control Number: 2021938016

ISBN: 978-1-63712-003-3 (paperback), 978-1-63712-005-7 (Kindle), 978-1-63712-004-0 (EPUB)

BISAC:

SCI027000 SCIENCE / Life Sciences / Evolution

SCI075000 SCIENCE / Philosophy & Social Aspects

SCI034000 SCIENCE / History

SCI015000 SCIENCE / Space Science / Cosmology

Publisher Information

Discovery Institute Press, 208 Columbia Street, Seattle, WA 98104

Internet: discoveryinstitutepress.com

Published in the United States of Ameria on acid-free paper.

First edition, first printing, August 2021.

ENDORSEMENTS

A brilliantly synoptic, dispassionate overview of the controversies that have swirled around Darwin's theory of evolutionary transformation over the past 160 years. The more that science has progressed, argues Neil Thomas, the greater the dissonance between Darwinism's simplistic mechanism and the inscrutable complexities of life it seeks to explain. Thomas's open-minded interrogation of the implications for our understanding of ourselves and our world is masterly and persuasive.

—James Le Fanu, Fellow of the Royal College of Physicians and winner of the *Los Angeles Times* Book Prize

Taking Leave of Darwin bristles with righteous indignation. Retired British humanities professor and lifelong rationalist Neil Thomas believed the confident claims for Darwinism. Now he knows better. Writing in elegant, erudite prose, Thomas excoriates those who have robbed people of their right to grapple with our mysterious universe as best they can. I highly recommend the book.

—Michael J. Behe, Lehigh University Professor of Biological Sciences and author of *Darwin's Black Box*

Professor Neil Thomas has written a brief, courageous, spirited, and lucid book. It shows the commendable willingness of a committed agnostic intellectual to change his mind about Darwinism, the great contemporary sacred cow, in the face of the large, accumulating body of new evidence against it and also to avail himself of the insights and arguments of intelligent critics of it since the very beginning and across 160 years—including Sedgwick, Mivart, Butler, A.R. Wallace, Agassiz, Max Muller, Kellogg, Dewar, Jacques Barzun, and Gertrude Himmelfarb. His intelligent, non-specialist survey of the contemporary state of the question is enriched by references to the insights of the distinguished philosopher Thomas Nagel and the MD and award-winning science writer James Le Fanu, and by a quite moving rationalist commitment to "follow the argument where it leads," however unexpected and uncomfortable this loyalty to logic and truth has made him. He provides a gratifying and illuminating case study in intellectual courage.

—M.D. Aeschliman, Professor Emeritus, Boston University, author
of *The Restoration of Man: C.S. Lewis and the Continuing Case
Against Scientism*

Taking Leave of Darwin provides helpful cultural and literary context for the development of Darwin's ideas and traces the rational and philosophical analyses that followed in its wake. Neil Thomas argues that even though Darwin's story of blind evolution is poorly grounded and inadequately supported, the public and some professional scientists (willfully or otherwise) have been duped into accepting a set of unsubstantiated assumptions and assertions. Despite being wary to entertain the notion of design, Thomas finds himself cornered—teleology is unavoidable.

—David Galloway, MD DSc FRCS FRCP FACS FACP, Former
President, Royal College of Physicians and Surgeons of Glasgow,
Honorary Professor of Surgery, College of Medical, Veterinary &
Life Sciences, University of Glasgow

Taking Leave of Darwin by Neil Thomas, a senior academic literary historian and a life member of the British Rationalist Association, is an unusual book, for it is an attack on the secular theory of evolution by a non-Christian. Thomas describes "the whole Darwinian edifice as an offense not only to best scientific practice but even to common sense." So why do Darwinists persist? According to Thomas, "If you bring to origins science an anything-but-God mindset, then you will cling tenaciously to the one purely materialistic theory you believe has any chance, however slender, of explaining the mysterious origin of life's diversity." Indeed, he says that "to attribute creative potential to nature itself is a deeply archaic, animistic way of thinking," and he concludes that "with the naturalistic/materialistic alternative having failed so signally, we are left with no other choice but to consider the possibility of the 'God hypothesis.'" I confess that I am not qualified to judge the book's scientific arguments, but am astonished by a non-Christian suggesting that Darwinism, far from a rigorous scientific theory, "cannot in strictly logical terms rise above the status of a hypothesis or philosophical postulate." If you are looking for a skeptical take on evolutionary theory outside the often simplistic science-vs.-religion framework, *Taking Leave of Darwin* is made to order.

—Rev. Dr. Paul Beasley-Murray, former principal of Spurgeon's College, London, and author of many books, including *There is Hope*

This well-researched and detailed examination of the all-pervasive and inexplicable overreach of neo-Darwinism in science and philosophy is a must read for all those interested in the science of origins and the nature of scientific conclusions. The analysis of historical objections to and current doubts about Darwinism are most illuminating. The author's reasons for "Taking Leave of Darwin" and being open to the existence of design in nature are compelling. A brilliant and thought-provoking book!

—Dr. Alastair Noble, Former Science Teacher and Inspector of Schools, Scotland, UK

Unlike many books written in the heat of the debates over Darwinism, *Taking Leave of Darwin* is a reflective, even meditative work. It charts the path by which one highly educated non-scientist has become an agnostic about the evolution of life by purely naturalistic processes. A thought-provoking feature of Neil Thomas's ruminations is the explicitness with which he considers how biology might have developed without Darwin's extreme naturalism. He concludes interestingly that it would have probably led to some sort of rapprochement with ideas of design and purpose in nature, perhaps along the lines of Alfred Russel Wallace's alternative treatment of natural selection. It remains an option worth exploring.

—Steve Fuller, University of Warwick, Author of *Science vs. Religion?* and *Dissent over Descent*

In a highly engaging and thoroughly researched account that is at once deeply personal and highly perceptive, scholar Neil Thomas exposes the bold presumptuousness of the Darwinian faithful for what it is, namely, a "modern form of hoodwinking" that fails an honest and unbiased test of history and logic. *Taking Leave of Darwin* exposes the "science" of Darwinian evolution as scientism and the explanatory "power" of natural selection as the paper tiger of "Nature red in tooth and claw." Read this book and be fooled no longer!

—Michael A. Flannery, Professor Emeritus, UAB Libraries, University of Alabama at Birmingham, author of *Nature's Prophet: Alfred Russel Wallace and His Evolution from Natural Selection to Natural Theology*

CONTENTS

PROLOGUE

WHAT IF CHARLES DARWIN GOT IT WRONG? WHAT IF ALL THE crises, alienations, and losses of faith we associate with the aftermath of the publication of *The Origin of Species*[1] had been triggered by a false prospectus? What if the latent but ever-present hostilities between science and religion of the last 160 years had been fomented by the equivalent of a "dodgy dossier"?

Like many others who "learned about" Darwin in school, I internalized his ascent-of-man narrative without demur, through what in retrospect seems like little more than a passive process of osmosis. By the second half of the twentieth century, Darwinism had become accepted as part and parcel of the mental furniture and indeed the fashionable thinking of the day, such that it would have seemed politically incorrect (and worse, un-hip) to challenge the truth-status of *The Origin of Species*. I must certainly have thought so since I recollect showing off my (superficial) knowledge of Darwinism to my first girlfriend, and doing so absolutely convinced that what I was saying was uncontestable.

To be sure, it had sometimes struck me that *The Origin of Species* contained some strange and counter-intuitive ideas, but I told myself that modern science is often counter-intuitive[2] (remembering the vast indeterminacies thrown up by recent advances in quantum theory), and I gave the matter little further thought. Since Darwin had been fêted by the scientific community for more than a century and a half, I deferred to what I imagined must be the properly peer-reviewed orthodoxy. Surely, I reasoned, any opposition to Darwin must be confined to the peripheral ranks of Biblical fundamentalists and young-earth creationists.

This complaisant (and complacent) stance was rather shaken when more recently I encountered some less easily disregarded opposition emerging from some of Darwin's latter-day peers in the ranks of scientific *academe*. Collectively, these publications made me alive to the possibility that the grand story of evolution by natural selection was little more than a creation myth to satisfy the modern age; and I found it impossible to ignore the dispute as being a "merely academic" issue, for if there is one subject which has had huge, often convulsive implications for the generality of humankind, it is Darwin's theory of evolution.

Any dispute concerning Darwin must necessarily have far-reaching implications beyond the guild of the biological sciences. It is not given to many to be able to muster the kind of equanimity shown by Charles James Fox Bunbury (brother-in-law to nineteenth-century geologist Sir Charles Lyell), who opined that, mortifying as the notion of human descent from jellyfish might be, "it will not make much difference practically"[3]; or to equal Dr. Johnson's priceless reaction to the notion held by an eccentric nobleman (Lord Monboddo) that man could be descended from apes: "Conjecture as to things useful, is good, but conjecture as to what would be useless to know, such as whether men went on all fours, is very idle."[4]

The majority of our Victorian forbears certainly could not find it within themselves to be so philosophical about a theory of human evolution that projected them into "a suddenly mechanistic world without a mechanic," to borrow a phrase used by Noel Annan in his biography of Sir Leslie Stephen (said to have lost his faith after reading Darwin).[5] This sense of being cast adrift from the erstwhile reassurances of the Christian faith was at painful variance with the paradigm of a providentially directed cosmos which had prevailed throughout the Christian centuries up to 1859.

In addition, when Darwin discharged his famous Parthian shot twelve years after publication of *The Origin of Species* in the *Descent of Man* (1871),[6] with its notorious claim of humankind's consanguinity with simian forbears, this amounted to a rather unambiguous demotion of humankind to a considerably lesser place in the scheme of things than its wonted pedestal just "a little lower than the angels,"[7] a demotion later exacerbated by Sigmund Freud's conclusions about the "hominid" nature of our subconscious minds.

It struck me that if a group of tenured academics and other responsible scientists could no longer support the claims on which these devastating inferences depended, and on which the worldview of much of the West presently rests, then this was surely a matter of some existential moment. Such disquieting possibilities drove me to investigate for myself the dispute between pro- and contra-Darwin factions. I make no apology for having made the attempt to read my way into a subject for which I have no formal qualifications, since my researches have led me to the conviction that the subject is of too universal an import to be left entirely in the hands of subject specialists, some of whom exhibit an alarming degree of bias and intransigent *parti pris* unconducive to the dispassionate sifting of scientific evidence.

Few coming to this subject can of course claim to occupy that fabled Archimedean vantage point of "seeing things clearly, and seeing them whole," and I make no such hyperbolic claim for myself. However, given the dismayingly sectarian nature of many evolution debates, it is a tedious but unavoidable necessity that I should add here at the outset that I have long been a non-theist and can at least give the assurance that the critique which follows will be based solely on rational criteria and principles.

The book is structured as follows. In the first chapter, I introduce the broad subject of how Charles Darwin and Alfred Russel Wallace came to formulate their theory of evolution by natural selection.

The second chapter looks at Darwin's intellectual formation from boyhood to maturity and the immediate reception of his *Origin of Species* with non-specialist British readers.

Chapter 3 turns to the mostly critical nineteenth-century reviews and receptions of *The Origin of Species* in the years and decades just after its publication, before Darwin had become the respected sage of his later years. The refreshing honesty of the early responses gives added clarity to the voices of dissent from Darwinism that were always present but which have become more insistent in recent decades. Those more recent responses are also covered in this chapter, together with the fraught issue of the fossil evidence marshalled to support Darwin's claims (which is exiguous and has occasionally even been proven fraudulent). We then look at what is in effect Darwin's companion volume to the *Origin*, namely *The Descent of Man*.

The fourth chapter considers those cosmological discoveries in the last half century with a bearing on the question of how the earth gained the unique supportive biosphere which enabled the evolution of plants, animals, and humans. Thereafter I unpack, and in some cases unmask, the frequently unacknowledged religious or anti-religious attitudes which have scarred the search for solidly based empirical findings for more than a century and a half.

In Chapter 5 I turn to the subject of what we can reasonably expect of the scientific method and what not to expect in the perennial quest to reveal the mysteries of life. In particular I question whether unrealistic expectations have led to questionable conclusions and issue an open invitation to subject specialists to reappraise the whole subject of natural selection as an evolutionary pathway.

In the final chapter I draw together threads from previous chapters to form a concluding synthesis. I round off the volume with some reflections on how researching and writing about this subject has brought me to a place I would have found surprising before I embarked on the project, especially regarding the intersection of science and religion. A short epilogue is also appended.

1. THE BATTLE IS JOINED

I would like to defend the untutored reaction of incredulity to the reductionist neo-Darwinian account of the origin and evolution of life. It is prima facie highly implausible that life as we know it is the result of a sequence of physical accidents together with the mechanism of natural selection.... I find this view antecedently unbelievable—a heroic triumph of ideological theory over common sense. The empirical evidence can be interpreted to accommodate different comprehensive theories but in this case the cost in conceptual and probabilistic contortions is prohibitive.

—Atheist philosopher Thomas Nagel[1]

In 1959, against the backdrop of a formal eulogy accompanying the centenary celebrations of the first edition of Charles Darwin's *The Origin of Species*, Julian Huxley (grandson of Darwin's great ally, Thomas Huxley) bestowed the status of *fact* on Darwin's theory of natural selection. But there were, and remain, well-informed holdouts, those who contend that Darwin's theory, despite influential supporters, cannot in strictly logical terms rise above the status of a hypothesis or philosophical postulate. Any theory about biological development claimed to have occurred millions of years in the past clearly cannot be tested by those conventional scientific procedures used to confirm or else disprove theories about contemporary phenomena. We cannot climb aboard the next available Tardis as time travelers to see for ourselves the truth or otherwise of Darwin's conviction that you and I ultimately evolved from microorganisms over long ages. Because of the limitations in testing a hypothesis about matters deep in the past, by 1897 the Oxford philosophy professor F. C. S. Schiller had gone on record to state that Darwin's idea was "a methodological device" rather than "an incontestable fact."[2]

There is, as well, the problem of contrary evidence—some of it stubbornly so. But before touching on that, it is only proper to acknowledge what should be obvious to anyone with even a nodding familiarity with evolutionary theory, namely that both Darwin's original theory and its various descendants (e.g., neo-Darwinism, punctuated equilibrium, the extended evolutionary synthesis) are hardly without evidential support.

The foundations of Darwin's theory of evolution by natural selection may be said to rest on expert observation (from his extensive fieldwork in South America aboard HMS *Beagle*) linked to intelligent speculation. His fundamental idea essentially concerns the inheritance of changed characteristics that provide superior adaptations to any given animal's environment. According to Darwin's supposition—taking the giraffe as an example—some individual giraffes were, completely by chance, born with slightly longer necks and so gained the selective advantage of being able to reach higher branches for food. This improved their chances of survival and of finding opportunities to mate and, hence, their chances of leaving offspring who would inherit longer necks. Repeated over generations, this would account for the superior length of their necks. Over time, the species that were best equipped by happy chance to survive would thrive and reproduce, while others would run the risk of becoming extinct.

This has remained the bedrock of the Darwinian understanding of evolution, even when the theory was later yoked to the emerging field of genetics. (And even still more recent versions of modern evolutionary theory that emphasize neutral evolution over natural selection continue, upon closer inspection, to allow a key role for natural selection at specific stages of the evolutionary process.)

As for the random half of his evolutionary equation, Darwin had been unable to suggest reasons for random variations in the first place. He believed they could be linked to random disturbances in the reproductive system, but it was purely speculation. Several critics pointed out that, as long as the origin of variation went unexplained, Darwin's evolutionary theory could not be accounted a complete mechanism of evolution. They noted that the initial variation was the truly creative force, selection merely eliminating those of its products which were substandard; and Darwin had little if any idea what that creative force was.

Darwin, so the story goes, was completed by Mendel. Darwin himself was unaware of the groundbreaking garden pea experiments conducted by the Moravian monk Gregor Mendel in the 1860s, the significance of which became widely appreciated only in the early part of the twentieth century. This eventually led to neo-Darwinism, where the findings of what is now called genetics were synthesized with the Darwinian model. Using standard terminology, one would say that natural selection is the process whereby persons carrying better genes in terms of survival and reproduction tend to have a larger number of offspring and that those offspring in turn are more likely to have better genes for survival and reproduction. Over generations the process becomes a virtuous genetic circle: beneficial mutations will prevail, harmful ones will tend to be eliminated, resulting in evolution toward better adaptation.

So the two parts of neo-Darwinian evolution are random genetic mutations and natural selection. Genetic mutations of all sorts occur, some destructive to the organism, some benign. Natural selection then sifts these mutations, eliminating some and preserving others. Many mutations are eliminated because they cannot survive in their natural environment, whereas others survive because they are better adapted to that environment. Mice that grow more fur in cold climates survive and therefore reproduce more abundantly than hairless mice, which are less likely to live long enough in the cold climate to reproduce at all.

Notice that on the Darwinian model, both in its original form and in the neo-Darwinian updating, neither the mice nor the environment set out to evolve furrier mice. The process is purposeless. Darwin repeatedly stressed this point. His evolutionary mechanism is blind and so heedless of the needs of the organism. He later conceded that "natural preservation" might have been a better term than "natural selection" with its misleading connotation of purpose. It was in fact entirely down to luck whether any change might occur which could confer a slight adaptive advantage.

The Malthus Turn

DARWIN'S THINKING was decisively stimulated by his reading of the *Essay on Human Population* (1798) by the economist and demographer Thomas Malthus.[3] Malthus's view of humankind was similar to, and in part influenced by, that pessimistic political philosopher who had lived a century before him, Thomas Hobbes. Hobbes saw men and women as selfish creatures, motivated only by personal advantage. In *Leviathan* he stated, "During the time men live without a common power to keep them all in awe, they are in that condition which is called war; and such a war as is of every man against every man"—the latter being the famous dictum of the *bellum omnium contra omnes*. If left in the state of nature, people's lives would be, wrote Hobbes in another phrase since become proverbial, "solitary, poore, nasty, brutish, and short."[4] One of Malthus's major themes concerned competition between humans for resources to survive, because he too felt that, left in a Hobbesian state of nature, each man's hand would be raised against the other, and the devil take the hindmost. He warned of the dangers of overpopulation, leading to the sobriquet of "Population Malthus" being widely applied to him. He was in fact the historical prototype for Charles Dickens's Scrooge figure with his dirgeful refrain of "overpopulation, overpopulation."

Reading Malthus by chance for purely recreational reasons, it suddenly dawned on Darwin how he might usefully appropriate the Malthusian analysis, essentially redirecting Malthus's ideas about the human struggle for existence, and overpopulation, to the wider biological world. From this, Darwin explains in his autobiography, he concluded that in the face of population growth, limited resources, and a struggle for survival that not all could win, "favourable variations would tend to be preserved, and unfavourable ones to be destroyed. The results of this would be the formation of a new species. Here, then, I had at last got a theory by which to work... I saw, on reading Malthus on population, that natural selection was the inevitable result of the rapid increase of all organic beings... [Malthus] gave me the long-sought clue to the effective agent in the evolution of organic species."[5]

Malthus's considerable influence on the development of evolutionary thought can also be found in the similar epiphany experienced by Alfred Russel Wallace, who came upon his ideas a decade or so after Darwin. Wallace, in his autobiography, also expresses his debt to Malthus for prompting him to a recognition of what he terms the "self-acting process" of evolution, and he expressed this discovery in terms which I believe make the issue particularly clear and which I cite here:

> One day something brought to my recollection Malthus' *Principle of Population*. I thought of his clear exposition of "the positive checks" to increase which keep down the population... why do some die and some live? And the answer was clearly, that on the whole the best fitted live. From the effects of disease the most healthy escaped; from enemies the strongest, the swiftest or the most cunning; from famine, the best hunters or those with the best digestion, and so on. Then it flashed upon me that this self-acting process would necessarily improve the race, because in every generation the inferior would inevitably be killed off and the superior would remain—that is, the fittest would survive... I had at last found the long-sought-for law of nature that solved the problem of the origin of species.[6]

This is of considerable historical interest. However, exactly how they got to the idea is less important than the question of whether the idea holds together. Darwin and Wallace each argued that the selection/preservation process could lead to the creation of new species. On a first reading, this might seem to be a non sequitur; but spelling out what they meant here, their presupposition was that successful members of any given species would develop to such an extent that they would become, over countless eons, superior forms unrecognizable as having sprung from the older, inferior biological stock.

Darwin introduces this thought in a somewhat unheralded way, but this in part was because he was using an elliptical shorthand meant primarily as an *aide-mémoire* to himself rather than the formal code of a public statement, where the missing logical link would have to be supplied. The apparent non sequitur, however, makes perfect sense within the context of the century of evolutionary theory which had preceded Darwin.

During that period a number of naturalists had mooted the possibility of one species modulating biologically into another one over vast swaths of time. This idea had become so common that it had long since been lexicalized in French in the term *transformisme* (associated primarily with the French naturalist Jean-Baptiste Lamarck) and, in English, "transmutation" (most closely associated with Darwin's own grandfather, Erasmus Darwin). It was clearly this tradition to which Darwin was alluding, a tradition which he may have thought did not need to be spelled out in greater detail.

He did offer physical evidence for his idea in *The Origin of Species,* arguably the most picturesque and memorable being the evidence derived from shared characteristics across disparate species. Darwin's *modus operandi* as a naturalist was, first, simple observation. He noted that there are certain undeniable similarities in the form and function among a human hand, a mole's paw, the leg of a horse, a porpoise's paddle, and the wing of a bat. They are constructed physiologically on the same pattern and with a comparable bone structure. For Darwin these likenesses seemed to point to a distant relationship and an inheritance from a common ancestor. In other words, if you take away the idea that all the species were created fully formed by some divine power, then the idea of descent from a common origin makes theoretical sense of these commonalities. Cumulatively, over "Jurassic" swaths of time and by an aggregation of small incremental differences, this process, Darwin argued, resulted in a thoroughgoing transmutation of species, starting from microscopic beginnings in the form of unicellular common ancestors, like bacteria, via numberless further stages up to ape-like intermediaries, thence on to the evolution of *homo sapiens.*

With Friends Like These

DARWIN FREQUENTLY conceded the evidential gaps and potential weaknesses of his theory, and referred to them in the conclusion of the *Origin*, where he wrote, "Any one whose disposition leads him to attach more weight to unexplained difficulties than to the explanation of a certain number of facts will certainly reject my theory."[7] Yet for Darwin the unexplained difficulties were outweighed by the explanatory power of his view. "I am fully convinced of the truth of the views given in this volume,"[8] he declared. Many others, however, disagreed with Darwin's speculation, and it was vigorously disputed by various eminent scientists in the decades following the publication of the *Origin*. These included some distinguished thinkers who were otherwise on very good, even friendly terms, with Darwin.

Asa Gray, the solicitous American natural science professor who frequently corresponded with Darwin, protested that evolution without divine design brought with it a host of problems.[9] Sir Charles Lyell, whose three volumes entitled *Principles of Geology* (1830–1833) inspired Darwin when he read the first of those volumes on his famous HMS *Beagle* voyage, had not reconciled himself to Darwin's theory even thirteen years after its publication. At that time he wrote that the basic problem of creation/evolution remained as inscrutable as ever despite what he diplomatically termed Darwin's impressive evidence. Essentially, he remained unmoved from his prior (theistic) opinion that the mystery of the origin of variations in the biological world involved "causes of so high and transcendent a nature that we may well despair of ever gaining more than a dim insight into them."[10]

Lyell did attempt to square Darwinian notions with his own beliefs, but only, it must be noted, at the cost of misrepresenting Darwin quite signally. Darwin had proposed a purposeless process; the evolutionary process Lyell described was shot through with teleology. "The more the idea of a slow and insensible change from lower to higher organisms, brought about in the course of millions of generations according to a preconceived plan, has become familiar to men's minds," Lyell wrote, "the more conscious they have become that the amount of power, wisdom, design and forethought required for such a gradual evolution of life, is as great as that which is implied by a multitude of separate, special and miraculous acts of creation."[11]

It is possible, if so minded (and Lyell and a number of other Victorians *were* so minded), to frame human evolution in progressionist terms as the ascent of man from inauspicious beginnings to ultimate human pre-eminence, superintended by a transcendent intelligence. This interpretation could then be used to facilitate a theistic accommodation with Darwinian gradualism, which, for all that it differed from the orthodox Genesis narrative, could still be glossed as an alternative providential route.[12]

Such, for instance, was the interpretation (some might argue rationalization) adopted by the clergyman author of *The Water Babies*, Charles Kingsley. In a letter to Darwin, Kingsley wrote that it was "just as noble a conception of Deity to believe that He created primal forms capable of self-development" as it was "to believe that He required a fresh act of intervention to supply the lacunas which He himself had made."[13] In place of the conception of a static creation—which for Kingsley conjured up disturbing notions of the indifferent *deus absconditus* [absent God] of the Deists—Kingsley rejoiced in a more dynamic, "hands-on" process benefiting from the *continuous* supervision of the Creator, which he took to be implicit in Darwin's narrative.

Darwin was not prepared to countenance such accommodations and bridled at such would-be Christian hijacking of his ideas, writing to Lyell: "I would give absolutely nothing for theory of nat. selection, if it require miraculous additions at any one stage of descent."[14] As Darwin saw the matter, the views of Gray, Lyell, and Kingsley, by espousing the higher law of providential arrangement, would have put biological science back into the realm of special creation, this being the very metaphysical quagmire from which Darwin was anxious to extricate it.

Darwin had especially hoped to convert Lyell to his viewpoint, since Lyell's theory of uniformitarianism in geology appeared on the face of it to cohere with Darwin's gradualistic ideas about human evolution. Along with Malthus, Lyell had been a major inspiration for the evolution of Darwin's thinking and also for Darwin's self-definition as a professional scientist, for Lyell represented the growing scientific trend towards wholly naturalistic, non-theistic explanations of terrestrial phenomena.

Lyell's geological positioning was conspicuously opposed to "Mosaic geology" and what was termed the doctrine of catastrophism. According to that hypothesis, the world arose through a sequence of catastrophes causing multiple extinctions, Noah's Flood being the last of these. Each catastrophe was redeemed by God through his creation of new species of ever greater complexity. The idea was that it had pleased God to improve on his Creation after the extinctions following each catastrophic event.

Lyell's opposing theory, uniformitarianism, was first proposed by the Scottish geologist and polymath James Hutton in his *Theory of the Earth* (1788), and later endorsed and developed by Lyell in his three magisterial volumes. Hutton did not think that catastrophes were necessary to account for terrestrial changes. Rather, he argued that they had been caused by the same forces as those active in his own day: frost, wind, running water, and the internal heat of the earth. Since such gradual, attritional forces would have required millions of years to achieve their effect, uniformitarianism stood squarely in opposition to popular biblical interpretations suggesting a much younger earth, such as that of seventeenth-century Bishop Ussher of Armagh, who computed its age as approximately six thousand years.

One can see why Darwin thought he might gain Lyell's endorsement for his purely naturalistic theory of humankind's development, but Lyell made a distinction in his own mind between the geological record and the world of organic and human life. Darwin was thus obliged to forgo the powerful endorsement of the renowned Lyell.

Doubters Near and Far

DARWIN FARED little better with other influential opinion-formers. His opponents included the Duke of Argyll, a Scottish nobleman with a serious and respected interest in naturalist ideas; St. George Mivart, a leading London science professor; and Richard Owen, England's leading paleontologist of the period. All of these vehemently contested the idea that natural selection could produce new species. Louis Agassiz too, the influential Swiss naturalist who had gained a chair at Harvard, contended that the beginnings of any given species were unobservable and beyond the capability of any science to explain. He insisted that to speculate without observable facts was neither responsible nor logically feasible.

It has become the watchword of modern science that large claims must be accompanied by large proofs, and surely one of the largest scientific claims ever made is the one Darwin made for his descent-with-modification theory of natural selection. Even now, more than a century and a half after the publication of his *Origin*, it has the power to shock and bewilder, especially when transposed from technical terms into the currency of everyday vernacular. A recent science writer, for instance, described natural selection in somewhat droll terms as "a mechanism powerful enough to turn fish into giraffes, given 400 million years."[15]

Nonetheless, my sense is that a good number of persons today, having noticed the conspicuous morphological likeness between man and ape, would entertain the supposition that *homo sapiens* evolved from ape-like ancestors, and suspend disbelief over the matter of the different cognitive capacities of man and ape. After all, unlike my nineteenth-century forbears, the great majority of my peers and I are not constrained by the biblical testimony that Adam was created by God in an instant, the divine breathing life into the dust of the Earth. This is a view many of us have lost with the progressive demythologization[16] of Scripture in the later nineteenth and twentieth centuries. We are the more prepared to accept an entirely natural explanation for the claimed ape-to-man evolution since we no longer have to agonize about where and when along the supposed evolutionary pathway any "ensoulment" might be imagined to have taken place.

But what of microbe to mammal? That the remote ancestors of apes and humans ultimately derived from minute bacterial forms would, I think, still cause many people to pause, even the most modern in outlook. For put in those terms, Darwinism sounds as weirdly incredible as anything found in Ovid's *Metamorphoses*, or the shape-shifting myths of pre-Christian Celtic tradition, or even Erich von Däniken's belief that transmutation of species had been caused by invading aliens manipulating the genetic codes of animal and human life.

But weirdly incredible or no, a development of the simplest prokaryotic life in the direction both of *homo sapiens* and gigantic animals such as elephants and the extinct dinosaurs is precisely what Darwin proposed and is still proposed by Darwinism's present-day intellectual descendants, as one of Darwin's most gifted expositors, Richard Dawkins, recently spelled out. "Natural selection happens naturally, all by itself, as the automatic consequence of which individuals survive long enough to reproduce, and which don't..." he writes. "Given enough generations, ancestors that look like newts can change into descendants that look like frogs. Given even more generations, ancestors that look like fish can change into descendants that look like monkeys. Given yet more generations, ancestors that look like bacteria can change into descendants that look like humans."[17]

That ambitious contention will not be left unexamined, and the whole issue of the tenability of the natural selection theory will be revisited below. For now, having set the scene for Darwin's intellectual journey, and before returning to his evolutionary ideas in more detail, I wish in the next chapter to turn to a brief biographical consideration of Darwin's earlier formation to try to assess how a less-than-stellar schoolboy and student came upon that theory which others had sought unsuccessfully for more than a century before his birth.

Also in the next chapter, attention will turn to a topic tactically skirted in *The Origin of Species* but which nevertheless provides the essential foundation and indeed precondition for that work, namely, the issue of the absolute origin of life on earth. For although Darwin said that solving the mystery of life's origin was *"ultra vires* [beyond our powers] in the present state of our knowledge,"[18] he still permitted himself to speculate privately in an earlier letter that it could have begun by spontaneous generation (i.e., not by divine Creation) from an accidental reaction of chemical elements within a "warm little pond."[19]

This postulate, first advanced by the ancient Greek Atomist philosophers (about whom more below), went on to underpin twentieth-century notions of "chemical self-assembly" in a watery medium, the process now termed "abiogenesis." At this pre-organic stage, as yet unknown chemical reactions are claimed to have produced rudimentary life forms, setting the stage for sentient life to eventually enter the scene. Those rudimentary life forms would, on that theory, become the putative raw material for natural selection to go to work on. The question of why Darwin may have elected to omit any formal consideration of this topic in the *Origin* also will be explored.

2. THE EVOLUTION OF A MYTH

Truly, scientific orthodoxies, like other orthodoxies, are sometimes very strange; and it is odd that scientists are so susceptible to self-hypnotic indoctrination.
—Cambridge entomologist William H. Thorpe[1]

"THE MOST DANGEROUS MAN IN ENGLAND." IN SUCH TERMS WAS Darwin described by a passer-by to his companion who spotted him by chance in London in 1863. Yet the naturalist who disturbed the settled belief-patterns of Victorian Britain and who gave rise to controversies which have endured to the present day was the very last person who would have harbored any desire to give offense to anyone. He appears, for instance, to have delayed publication of the *Origin* partly out of deference to the religious sensibilities of his wife, Emma. Darwin's life and personal formation reveal a decent, often retiring man, beset in his middle years by trying internal maladies, who had little of the firebrand or iconoclast about him.

The young Charles had the good fortune to have been born into an accomplished and well-to-do family. One of his grandfathers was the polymath Erasmus Darwin, who was both a medical practitioner and a very original combination of poet and evolutionary philosopher. His other grandfather was Josiah Wedgwood, the highly successful founder of the Wedgwood pottery enterprise. The Wedgwood manufacturing wealth, amplified by Darwin's father's twin career as both a medical doctor and financier, gave the family financial security, and grandfather Erasmus's scientific interests gave the young Charles an important intellectual hinterland which was to play a role in his adult career as a gentleman-naturalist. (Being independently well-off, Darwin never required, or had, a salaried job.)

Despite the many advantages heaped upon the young Charles by his privileged background, however, his early life and career were marked by indecision and lack of clear motivation. Furthermore, in purely scholastic terms, he evidenced little of the brilliance one might expect of one who was to become such a pioneer in the world of biological science. Darwin himself, unassuming and admirably "grounded" in his own self-assessments, would have been more than happy to concur with that verdict since his autobiography abounds in self-effacing anecdotes such as his being "slower in learning" than his sister, Catherine, or his difficulty in learning foreign languages and mastering the art of classical verse-composition when in school. In sum, he wrote, "When I left the school I was for my age neither high nor low in it; and I believe that I was considered by all my masters and by my father as a very ordinary boy, rather below the common standard in intellect."[2]

From 1818 to 1825 Charles attended Shrewsbury School, whose institutional rigidities he frequently sought to escape by looking for natural history specimens (his true interest) around and about the school's grounds. In 1825 his physician father dispatched him to the prestigious Edinburgh Medical School. Here Charles was able to pursue his natural history interests but, despite developing a wide range of scientific ideas in Edinburgh, he did not complete his medical course, leaving prematurely in 1827. He cited personal squeamishness about having to witness harrowing operations (this was an era before anesthetics), but it is difficult not to imagine that some maverick element of his personal formation played a part in his decision.

In addition, as his biographers Adrian Desmond and James Moore point out, his comfortable circumstances did not give him adequate motivation to concentrate his mind on success in a rather exacting professional discipline.[3] Darwin in fact admits as much in his autobiography: "I became convinced from various small circumstances that my father would leave me property enough to subsist on with some comfort, though I never imagined that I would be so rich a man as I am; but my belief was sufficient to check any strenuous effort to learn medicine."[4]

Subsequently, his father, anxious that his son should not become "an idle sporting man" (a propensity which Charles was inclined to indulge in the company of sundry "hearties" in the numerous homosocial groups encouraged by a then almost wholly male higher education sector), directed him towards Cambridge University. Here, in an age when no profound spiritual convictions were required for such a career, his son was set to study theology so as to attain that popular nineteenth-century fallback position of the English middle classes, that of finding a living as a parson. Yet Charles eventually declined this career path as well.

Not surprisingly at this point, there was increasing family anxiety that Charles might fall into that type of gentleman-idler which his material security would have permitted; but then his father (forced once more into the breach in an era when nepotism was not seen as problematical either by university authorities or in other walks of life) procured for him the role of gentleman companion to the ship's captain on HMS *Beagle*. It was in fact Charles's minor public school background (what in the United States is quite reasonably referred to as a "private school background") which commended him to the *Beagle's* captain, Robert Fitzroy—that and, we may surmise, the fact that Charles's father would be paying his son's expenses.[5]

That famous ship and its five-year voyage of exploration and observation took Darwin to South America and the South Sea Islands where he was able to catalog and study the flora and fauna of the region. It proved to be the making of the young naturalist, giving him more than ample opportunities for that protracted field work which formed the basis of his later writings. In short, this richly formative experience gave firm and sustained direction to the rest of his life, for it was in that remote and exotic milieu that Darwin, previous vacillations banished, was at last able to find himself.

After Paley

A COMMON trope of fictional books and films representing life in the earlier Victorian era is the eccentric-looking clergyman with a net on a long wooden handle used to collect butterflies and other wildlife from his countryside forays. The image so beloved of filmmakers is not just a picturesque fictional prop; it reveals what in many cases was an important reality of scientific endeavor in the first half of the nineteenth century, where naturalists, frequently men of the cloth, sought to analyze findings culled from nature in order to illustrate and illuminate what they firmly believed was the divine order of things. The assertion of Francis Bacon that God had authored two books, the Bible being one and the other inscribed in the very fabric of nature itself, had a long pedigree and lasted well into the high Victorian era, so much so that many scientists believed that it was the theological relevance of their searches which was the ultimate justification of their pursuit. Science supported religion and vice versa.

The doctrine that the design of nature permitted inferences about the nature of God himself was most influentially articulated in a book Darwin knew well and admired, William Paley's *Natural Theology* (1802, frequently reprinted throughout the nineteenth century).[6] Hence the harmonious adjustment of animals to their environment was glossed as an indication of God's providential arrangements: the wings of birds, for instance, clearly so vital to avian life, were interpreted as evidence of God's benign superintendence. Perhaps Paley's most famous illustration of divine intervention was his oft-cited watch analogy, which I will reproduce *in extenso* here both because it is the classic exposition of the belief but also because it provides some of the underpinnings of the intellectual structure of what is termed the modern intelligent design movement (about which more below):

> In crossing a heath, suppose I pitched my foot against a stone, and were asked how the stone came to be there, I might possibly answer, that, for anything I knew to the contrary, it had lain there for ever: nor would it, perhaps, be very easy to shew the absurdity of this answer. But suppose I found a *watch* upon the ground, and it should be inquired how the watch happened to be in that place, I should hardly think of the answer which I had before given, that, for anything I knew, the watch might

have always been there. Yet why should not this answer serve for the watch as well as for the stone? Why is it not as admissible in the second case as in the first? For this reason, and for no other, viz. that, when we come to inspect the watch, we perceive (what we could not discover in the stone) that its several parts are framed and put together for a purpose, e.g. that they are so formed and adjusted as to produce motion, and that motion so regulated as to point out the hour of the day; that, if the different parts had been differently shaped from what they are, of a different size from what they are, or placed after any other manner, or in any other order, than that in which they are placed, either no motion at all would have been carried on in the machine, or none which would have answered the use, that is now served by it.

To reckon up a few of the plainest of these parts, and of their offices, all tending to one result:—We see a cylindrical box containing a coiled elastic spring, which, by its endeavour to relax itself, turns round the box. We next observe a flexible chain (artificially wrought for the sake of flexure) communicating the action of the spring from the box to the fusee. We then find a series of wheels, the teeth of which catch in, and apply to, each other, conducting the motion from the fusee to the balance, and from the balance to the pointer, and, at the same time, by the size and shape of those wheels, so regulating that motion as to terminate in causing an index, by an equable and measured progression, to pass over a given space in a given time.

We take notice that the wheels are made of brass, in order to keep them from rust; the springs of steel, no other metal being so elastic; that over the face of the watch there is placed a glass, a material employed in no other part of the work, but in the room of which, if there had been any other than a transparent substance, the hour could not be seen without opening the case.

This mechanism being observed (it requires indeed an examination of the instrument, and perhaps some previous knowledge of the subject, to perceive and understand it; but being once, as we have said, observed and understood), the inference, we think, is inevitable; that the watch must have had a maker; that there must have existed, at some time, and at some place or other, an artificer or artificers who formed it for the purpose which we find it actually to answer; who comprehended its construction, and designed its use.[7]

The argument from design, which Immanuel Kant called the physi-co-theological argument, held the field largely unopposed until 1859, excepting the eighteenth-century rationalist philosopher David Hume's prior objection that the apparent design of nature did not logically per-mit any conclusion about its cause. In Hume's *Dialogues Concerning Natural Religion* [1779], Hume's skeptical character Philo argues for the uncertainty of design arguments by showing that the world is at least as much like a generated animal as like a designed machine:

> The world, say I, resembles an animal, therefore it is an animal, there-fore it arose from generation. The steps, I confess, are wide; yet there is some small appearance of analogy in each step. The world, says Cleanthes, resembles a machine, therefore it is a machine, therefore it arose from design. The steps here are equally wide, and the analogy less striking. And if he pretends... to infer design or reason from the great principle of generation... I may, with better authority... infer a divine generation or theogony from his principle of reason. I have at least some faint shadow of experience [i.e., empirical evidence]... Rea-son, in innumerable instances, is observed to arise from the principle of generation, and never to arise from any other principle.[8]

After more arguments showing the non-demonstrative character of design inferences, Philo concludes, "A total suspense of judgment is here our only reasonable resource."[9]

Such arguments from Hume probably helped weaken belief in tra-ditional religion, but it was only after the publication of Darwin's *The Origin of Species* that the antagonism between religion and science fa-miliar to us moved to the fore of society. That said, it should at the same time be added that Darwin represented a tipping point amidst a mood of unease which had been gathering speed for some decades. In 1831 John Stuart Mill concluded that the present era, where the predictable continuities of old agrarian economies had been increasingly ousted by the unlovely incursions of industrialization, was one in which people had "outgrown old institutions and old doctrines, and have not yet acquired new ones." Both Matthew Arnold's "Dover Beach," where the poet fa-mously hears "the melancholy, long withdrawing roar" of "the sea of faith,"[10] and Tennyson's *In Memoriam*, a poeticized funeral oration on the unexpected loss of his friend, Arthur Hallam, which leads the poet to "falter where I firmly trod" on the "world's altar stairs/That slope thro' darkness up to God," were written some years before 1859.

Tennyson's poem has even been adduced as a harbinger of the theory of natural selection, where the poet describes Nature "red in tooth and claw" as being "so careless of the single life." The savage voice of nature in that poem insists that it's even worse—not just individual lives but thousands of species wiped out, and the holy spirit reduced to mere "breath," and man to "desert dust."[11] Those particular lyrics were probably influenced by Tennyson's reading of Lyell's *Principles of Geology*, where he will have encountered the author's discussions of the extinction of species throughout the earth's history as they found themselves unable to cope with their environments.[12]

It was clear that Darwin's theories were inimical to a literal reading of the biblical creation narrative, but by the 1860s by no means all were biblical literalists: Lyell's findings from three decades earlier were widely seen in educated circles as having all but disproved the biblical flood narrative, at least when interpreted as a global flood.

As an aside, it is worth noting that even in the medieval and early modern periods there had been a tradition of a nuanced reading of the Bible on a number of levels beside the literal (which together were termed the four-fold exegesis). As David Knight has pointed out, "While our ancestors took it for granted that the Bible was the word of God, and literally true, they read or heard much of it as they would a great poem or play to be mulled over at more than one level."[13] Indeed, some of the Church Fathers did not view the days in Genesis 1 as twenty-four hour days. That is, they argued that it was intended to be read figuratively.[14]

Public Outcry

But now there appeared a more insidious threat which began to agitate people more than any question about the precise truth-status of Genesis, and this was the sense of an encroachment of science into areas heretofore reserved for theology: the mind and soul.

While science had eroded some theological territory before this time, by the 1860s its increasing threat grew into a kind of pincer movement that included the influence of the biblical "Higher Criticism" from Germany (which essentially submitted sacred texts to the same kind of dispassionate, forensic evaluation as secular ones). This trend was exemplified by the South African Bishop Colenso's *The Pentateuch and the Book of Joshua Critically Examined* (7 volumes, 1862–1875), or the fearlessly demythologizing *Essays and Reviews* authored by six ultra-liberal churchmen (1860), which treated the Bible essentially like a secular text. David Strauss's *Life of Jesus*, translated into English by George Eliot (*nom de plume* of Mary Ann Evans, 1846), which emphasized Christ's humanity rather more than his divinity, was another influential publication in the same vein. Many British readers failed to appreciate the semantic halo customarily attaching in its native country to the term *wissenschaftlich* (scientific) when applied to cold-eyed analyses of their King James Bible, and indeed the unfortunate Colenso ended up being arraigned for sacrilege in the ecclesiastical courts.

Not unexpectedly, then, there was no small measure of popular resistance to Darwinian notions. A forum for scientific discussion about the alarming new subject of creation and evolution, the Victoria Institute,[15] was formed in 1865, and its influence and activities endured well into the first half of the twentieth century. The Institute was rebranded in 1932 as the Evolution Protest Movement, which issued the following statement of purpose:

> We feel the public are being deceived. Evolution propaganda does not present the facts impartially; it dwells upon those which favour the theory, while suppressing those which oppose it. Such are not the methods of true, but of false, science. Few people realise that the tactics which Evolution employs would be regarded as "special pleading" in a Court of Law; and that many scientists have declared that Evolution is both unproved and unprovable.[16]

Douglas Dewar, the scientific brains behind the reconstituted organization, published two serious technical works which strove to point up the "extraordinary fallacies" undergirding Darwinism.[17] Resistance was also apparent in contemporary newspaper reporting, as Alvar Ellegård documented.[18] Here it is noteworthy that a good deal of the opposition came not from wounded religious sensibilities but from common-sense objections arising from people's instinctive trust in everyday forms of logic. Ellegård reports that Darwin's theory of the survival of the fittest, with its picture of creatures constantly strained by overpopulation and sifted by an unending existential struggle, was commonly rejected on the basis that, more often than not, the habitat in which animals have been placed gives them a sufficiency of resources.[19] Against the Darwinian claim that over "deep time" there must have been manifold instances of evolution of one species into another, it was argued that there was "no evidence... that there has ever existed a tendency, in a creature fitted for one sphere, to usurp that of another."[20] People could accept that the weak might be weeded out without accepting that the fit would necessarily get fitter: there was little acceptance that the process of natural selection could be creative. The whole descent-with-modification theory of animal metamorphosis was widely rejected for being "imaginary," especially since readers had noted that Darwin himself admitted that the fossil evidence was simply not there (yet) to support his claims.

Although people were inclined to accept the notion of evolution—which had already gained currency in various forms well before Darwin—they balked at the idea of evolution via chance variations, whether or not sifted by natural selection. "Things do not happen by chance" was the frequent riposte: forethought, planning, design, would have been necessary to effect such momentous changes as Darwin proposed.

Towards the end of the century such skeptics were to find an articulate champion in the unlikely form of novelist Samuel Butler, little remembered today except by students of literature (and then probably only in the context of "survey" courses). He is best known as the novelist who wrote *Erewhon* and *The Way of All Flesh*, satirical indictments of Victorian England's major institutions—the family, the Church, and snobbishly hierarchical class structure. Butler was an early supporter of Darwin, one whose early enthusiasm turned to disenchantment.

A complex and latterly somewhat disturbed personality, he spent a full decade of his life researching the subject of evolution. He had no scientific credentials for this task beyond a stint of sheep farming in New Zealand in the 1860s but applied himself with such assiduity to the task of debunking Darwin (by whom he had felt slighted over an intellectual copyright issue) that he became shunned by polite society and ostracized by those of the Darwin party. Hoping to find a refuge in Science from his first spiritual home, the Church, he found to his cost that, as Malcolm Muggeridge put it, "Science could be as dogmatic as any Church, and with less justification, and its devotees as bigoted as any country clergyman."[21]

Among the more venial faults he arraigned Darwin for was plagiarism[22]—for which there is some basis: Darwin's biographers consistently report that he was little inclined to acknowledge his intellectual debts, excepting the fulsome praise he gave to Malthus and Lyell. To Darwin's credit, he responded to such criticism by adding a "Historical Sketch" to the third edition of *Origin* (1861). There he listed thirty individuals he thought significant contributors to evolutionary theory.[23]

More substantially, Butler found the very mechanism of natural selection unconvincing. In his book *Luck or Cunning as the Main Means of Organic Modification,* he went into exhaustive detail on this issue, at times indulging in some rather monomaniacal streams of consciousness, bearing witness to the obsession the subject of evolution had become for him. For Butler, Darwin had muddied the waters of the heretofore teleological territory of evolution by what he saw as the illogical and distinctly un-teleological postulate of natural selection.

Although Butler had lost his Christian faith, he still retained a sense of what we might now term "spirituality" or, as his modern biographer noted, a belief in vitalism: "So he substituted for the exploded idea of instant, once-for-all creation a belief in the essential unity of life, life with a sense of will, purpose and progress."[24] Undeterred by any Arnoldian or Tennysonian mawkishness, Butler would appear, in company with many later Victorians, including William Hale White's semi-fictional figure of "Mark Rutherford,"[25] to have taken Wordsworth as his guide. In that poet's famous "pantheistic" lines in "Tintern Abbey," he reports being able to see into the life of things. This comes in a passage suggestive of the kind of divine immanence Butler envisaged:

> And I have felt
> A presence that disturbs me with the joy
> Of elevated thoughts; a sense sublime
> Of something far more deeply interfused,
> Whose dwelling is the light of setting suns,
> And the round ocean and the living air,
> And the blue sky, and in the mind of man;
> A motion and a spirit, that impels
> All thinking things, all objects of all thought,
> And rolls through all things.[26]

Butler denounced adherents of the Darwinian theory for being "apostles of luck." For him organic evolution depended not on luck but on "cunning"—a word he used in an idiolectal sense denoting something akin to a preconceived, thought-out natural law animating all nature. In this, Butler was giving his accomplished writer's voice to opinions circulating in the press and on the streets a few decades earlier, indicating that Darwin's attempt to delete any animistic notion from his wholly material theory of natural selection frequently fell on stony ground. Most readers thought some "law of development" (what Henri Bergson, at a later date, would term the *élan vital*) must have been implanted by the Creator as the necessary motor of evolution.

An Ancient Debate

THOSE LOOKING into the subject of creation and evolution may at first be somewhat surprised to find the names of pre-Christian Greek and Roman authors liberally referenced even in up-to-date discussions of the subject. It may at first appear incongruous that the names of philosophers belonging to what we now deem a largely pre-scientific era should be thought to have anything to contribute to the subject. However, when it comes to the topics of creation and evolution, there is little if any new under the sun. Those grand existential themes have been the subject of human speculation for millennia, and because the modern scientific method can take us only a limited way to understanding these eternal mysteries of the human condition, even modern scientists are thrown back on the resources of intelligent speculation to fill the gaps that "hard" science cannot answer. Whereupon they find that our ancient predecessors were there before them.

Thus, are we able to view Darwin's evolutionary ideas against the background of a preoccupation beginning some six centuries before the birth of Christ and continuing (albeit with large historical interruptions) to the time of Charles's grandfather and thence to prominent nineteenth-century near-contemporaries of Charles himself.

Such a contextualization seems particularly important in Darwin's case, since his *Origin of Species*, while incontestably based on the precise empirical evidence of many years of minutely observed fieldwork, had a large dimension of "natural philosophy" to it. Especially in the context of the nineteenth-century amateur-naturalist tradition to which Darwin belonged, this meant essentially the application of intelligent speculation to the putative mainsprings of natural phenomena. In addition, although Charles himself chose not to flag up his debt, he was clearly influenced by the ideas and writings of his grandfather. Some of Erasmus Darwin's books have been found to have his grandson's signature in them and to have internal written markings. Darwin freely admitted to have read his grandfather's *Zoonomia*, and few historians believe Charles's claim that it had no effect on him. See, for example, Loren Eiseley's *Darwin's Century*.[27] Charles claimed in his *Autobiography* that during his youthful walks with the materialistic evolutionist Robert Edmond Grant in Edinburgh, evolutionary concepts left little impression on him. More than one historian has found this claim disingenuous. As Janet Browne notes:

He [Charles] had by then studied Erasmus Darwin's evolutionary works, particularly *Zoonomia*. A previously unknown list made by Darwin of the books he read during his second year at Edinburgh makes it plain that he studied his grandfather's volumes closely—closely enough to read Anna Seward's biography of him (published in 1804).... Young Darwin, it now turns out, was well aware of evolutionary views and perfectly capable of grasping the full implications of what Grant had to say.[28]

This, and other evidence, has led historian Michael Flannery to conclude that as regards Darwin developing his own brand of evolution, "by the time Darwin finally put pen to paper his metaphysic had run well ahead of his theory."[29]

Not only did Dr. Erasmus inaugurate what has been jocularly termed the Darwinian cottage industry of evolutionary speculation, but he also (being unarguably more gifted than his grandson in purely scholastic terms) was well acquainted with the classical works of natural philosophy, on whose conceptions he based some of his own verse poems, as will be observed below.

Let us, however, begin at the beginning. Surviving written records indicate that philosophical speculation about the origin and development of the world reach back at least as far as the Greek Anaximander (611–547 BC) and his follower Anaximenes (588–542 BC), who thought that the earth was initially muddy and that out of this primordial slime there arose first plants and animals, then human beings. At first partly aquatic, humans subsequently moved their abode to land. Notably, Anaximander's wholly naturalistic explanation of things did away with the necessity for invoking mythological explanations involving the Greek gods.[30] Common to both ancient and modern debates about creation and evolution is a tension between the argument for divine creation-cum-superintendence and the opposing argument which strives to exclude god(s) and seek explanations for the phenomena of life along strictly material lines.

In the Homeric and Virgilian epics and in other imaginative literature of the Greek and Roman worlds, the numerous deities (often personifications of natural forces) appear in directly interventionist roles, but their existence was vehemently disputed in a number of ancient philosophical traditions. For philosophers such as Empedocles, Democritus, Epicurus, and Lucretius, life is not a divine creation but simply an emanation of the natural flux of things, part of a common continuum with the sea and sky. Empedocles addressed the problem of the world's complexity by speculating that the flux tossed up all sorts of different shapes and objects generated at random by the chance interaction of elements. One text above all others from Roman antiquity appears to have exerted a particular influence on the post-1700 world: *De Rerum Natura* (*On the Nature of Things*) by the philosopher-poet Lucretius (c. 50 BC). This work also influenced more than one generation of the Darwin family, as will become clear below.

The ultimate inspiration for Lucretius's extended verse poem was a philosophical treatise, *The Art of Happiness* by the Greek Epicurus (342–270 BC), whose austere propositions were transposed by Lucretius into a more accessible verse form, which enabled it eventually to capture the imagination of European posterity. What was the essence of the Epicurean philosophy versified by Epicurus's Roman disciple, and what is its relevance to the Darwins? Let us begin with its fundamental propositions, which I will summarize in the following three paragraphs.

The universe, according to Epicureanism, is mindless and without creator, being a purposeless and non-intelligent concourse of atoms without any cosmic source of direction sustaining it. Its invisible particles or atoms are constantly in motion, jostling against one another without guidance or direction. There is no end or purpose to existence, only ceaseless mutation, creation and destruction, governed entirely by chance, in which atoms swerve around now this way, now that. Since there is no original scene of mythic creation to be invoked, Epicureanism proposes that plants and animals evolved via an extended process of trial and error. This random process, which continued over immeasurable tracts of time, is said to be responsible for the emergence of all species, animal and human. In some cases that random process was unsuccessful, resulting in creatures not properly equipped to compete for resources or to create offspring, and which succumbed to extinction—in contradistinction to perfectly formed creatures able to adapt and reproduce.

In sum, they held that despite appearances to the contrary, things come about by happenstance rather than by design: sight did not exist before the birth of the eyes nor speech before that of the tongue (i.e., these organs were not created purposefully for our use). Language was not a divine gift. Humans, like animals, produced sounds, but in the human lineage those sounds in time evolved into more complex codes of understanding (although the precise mechanics of this human advance in sophistication went unexplained). Music was developed by humans imitating the warbling of birds. The earth was not created for human habitation, and it is a delusion to suppose we have a central position in it: there is in fact no reason to give humans a status greater than other animals with which they share many similar qualities. Humans are also part of a larger material process which links them not only to inorganic matter and the animal world, but even to the stars in the sky.

The origins of humans did not occur in some paradisal location but in a primitive battle of survival (of the fittest), struggling to eat and to avoid being eaten—although some rudimentary capacity for communal living did at length evolve. There is no soul and no afterlife, according to the Epicureans, and nobody should be concerned about his or her death since neither a paradisal nor an infernal fate awaits us at the end of our days. Indeed, there is no need to believe in any of the superstitious delusions promulgated by religion. People's fantasies about superior beings in the heavens who must be propitiated are without foundation. There are no Fates, harpies, daemons, genii, satyrs, dryads, and the like. Such delusions are simply obstacles to our happiness. Epicurus and his followers exhort us to forsake the cruelties of religion, which demand ascetic self-denial, violent retribution, and (in the classical world) even human sacrifice (as in Agamemnon's sacrifice of his daughter, Iphigenia). Bearing in mind that the Greek pantheon of gods was nothing if not fractious and homicidal, it is not difficult to see why the Epicureans would have viewed emancipation from such harmful fancies as conferring happiness on humankind.

The above summary unmistakably contains prototypical expressions of ideas favored by Charles Darwin. As Neal Gillespie put it, Darwin's "vision of a masterless and undesigned nature brought with it hints of ancient atomism and its attendant atheism."[31] The trial-and-error development of life described in Lucretius foreshadows in some sense the notion of natural selection, while the idea of animals too weak or ill-adapted to compete with their fellows brings to mind the Malthusian/Darwinian idea of the survival of the fittest. It is surely no mere coincidence that Renaissance scholar Stephen Greenblatt reports that he experienced a shock of recognition when, as a young student, he first encountered Lucretius's writings.[32] He was struck by the familiar atheistic tenor to which he had been exposed in the elite intellectual circles in which he moved in late twentieth-century America, circles which by that time had been adjusting to the influence of Darwinism for more than a century.

The ancient Greeks were not experimental scientists but thinkers who brooked no constraints on their speculative flights. As George Strodach explains, the atomists were good at producing "bold metaphysical postulates," but "the Greeks neither understood nor employed experimental method to any significant extent. In certain cases they erected brilliant hypotheses, such as the atomic theory, and then dogmatically asserted the truth of such hypotheses without rigorous testing."[33]

Whatever reservations might be held about the truth status of their ideas, however, it was more the sacrilegious nature of the Epicurean/Lucretian take on the world that proved so unacceptable to both ancient and medieval people. Hence atomism as a theory of reality swiftly disappeared from view and was almost lost to history. It was snatched from oblivion only when the humanist scholar and ancient manuscript-seeker Poggio Braccioloni tracked down a transcribed copy of it in a German monastery in the first part of the fifteenth century, and it was still not much visible until the seventeenth century, when it was set before the public eye by the Jesuit Pierre Gassendi, a contemporary of Descartes.

The poem's first translation into English came in 1682 from the pen of a young Oxford don, Thomas Creech, and this was republished throughout the eighteenth century. Hence its reintroduction into the European literary/philosophical canon came at a propitious moment coinciding with the beginning of the Enlightenment. By the end of the eighteenth century there is evidence that atomist ideas influenced David Hume in his *Dialogues concerning Natural Religion*, where one of the disputants in the imagined debate tells us that, over vast swaths of time, matter itself can produce ordered forms having the appearance of design. God's design, on that argument, represents an unnecessary hypothesis.[34]

Although there was a growing tolerance for the content of the *De Rerum Natura* near the end of the eighteenth century, the whiff of brimstone that persisted around it meant that Sir Isaac Newton, when he was falsely accused of "being a Lucretian," felt compelled to emphasize publicly that he did not embrace the atheistic doctrine.

Darwin's Modern Forerunners

CHARLES DARWIN's grandfather, Erasmus Darwin, was an open admirer of Lucretius as a poet and in part calqued his own natural science poems on the Lucretian model, as one contemporary critic noted when he wrote, "Dr Darwin, like Lucretius, has endeavoured to blend in his poetical works the grave features of philosophy with the mutable graces and smiling charms of imagination."[35] However, it is noteworthy that Charles's grandfather, while admiring Lucretius as a poet, chose to dissociate himself from the Lucretian philosophy, here expressed in unambiguous (but possibly at the same time also defensive) terms through the medium of one of his odes:

> Dull atheist, could a giddy dance
> Of atoms lawlessly hurl'd
> Construct so wonderful, so wise,
> So harmonised a world?[36]

This and other lines of evidence suggest that Erasmus Darwin remained a theist, even if he didn't believe in divine revelation. However, it must be added that one of Charles's biographers, Janet Browne, has reservations about the strength of the grandfather's faith, since the Darwin family were widely known to be freethinkers and his philosophical conjectures as expressed in his poetry were nothing if not *outré*. His *Temple of Nature*, for instance, mooted the possibility of life having emerged from the depths of the oceans and evolving into different species in response to a striving for perfection in different environments (the idea most closely identified with Lamarck in the nineteenth century).

It is also perhaps telling that initially Erasmus had doubts about publishing his poem *Zoonomia or the Laws of Organic Life* because he feared accusations of heresy, and, indeed, when it was published, it acquired the distinction of being banned by the Pope. Be that as it may, his conception of the beginning of the world and the subsequent evolution of its denizens was ostensibly "sound" theologically, as he sought to show in these oft-cited words:

> Would it be too bold to imagine, that in the great length of time, since the earth began to exist, perhaps millions of ages before the commencement of the history of mankind, would it be too bold to imagine, that all warm-blooded animals have arisen from one living filament, which THE GREAT FIRST CAUSE endued with animality, with the power of acquiring new parts, attended with new propensities, directed by irritations, sensations, volitions, and associations; and thus possessing the faculty of continuing to improve by its own inherent activity, and of delivering down those improvements by generation to its posterity, world without end![37]

What precisely Erasmus Darwin meant by the term "filament" is unknown, but it is generally held that he was referring (in modern terms) to a very small, possibly unicellular entity which thereafter branched out and developed into a series of larger species. It could be debated whether he viewed this process as unguided. The phrasing in the above passage suggests as much. On the other hand, he may have viewed this evolutionary process as taking place under God's continuing superintendence, since Erasmus did not subscribe, officially at any rate, to the "absentee landlord" conceptions of the eighteenth-century Deist creed. His grandson's theory of natural selection, on the other hand, as the term implies, was framed to explain human evolution in wholly naturalistic terms.

As our brief historical sketch draws closer in time to Charles's Darwin's revolutionary contribution to evolutionary theory, one thing to note is that he was anything but innocent of the battles royal that had been raging on the subject for the better part of a century. He was fully aware that his *Origin of Species* was part of a larger corpus of evolutionary thought, ancient but also modern.

Consideration was given above to the classical world's contributions to this perennial philosophical debate, but it was in the eighteenth century that modern ideas of evolution began to take shape.

The father of the modern discipline of biological taxonomy was the Swedish botanist Carl Linnaeus (1707–1778), whose opinion was that species do not change and that all living things were created as they can be observed today. On that basis he classified them into fixed groups, which he identified with the descendants of the original forms made by the Creator. His work was highly influential, to the extent that Linnaean categories are still referred to with profit today by animal taxonomists.

In the footsteps of Linnaeus came Georges Cuvier (1769–1832), whose numerous areas of expertise included zoology, comparative anatomy, and paleontology. Cuvier discovered that the fossil vertebrates could be placed in a sequence from fish to mammal, but he did not conclude that the sequence indicated that one form had descended from another. Rather was it the result of a succession of separate creations. Another French scientist, Georges Leclerc Buffon (1707–1788), with a stupendous forty-four volume *Histoire Naturelle* to his name, was an early proponent of a form of evolution, challenging the fixed categories of the Linnaean system of classification. Buffon held that living things evolved and that species would advance or regress as their environment changed, but, unlike Charles Darwin in the next century, he did not specify how these changes might occur, nor did he propose that the evolution could generate entirely new body plans.

A notable follower of Buffon was Jean-Baptiste Lamarck (1744–1829), who advanced the idea that some organisms might have developed from previous ones. He even put forth the idea that an animal could develop new organs in response to its need to operate in a changing environment. Such newly acquired traits would then be inherited and so contribute to the development and evolution of a species (an idea rejected by the scientific establishment of the day and by Charles Darwin, who, however, later in life expressed some sympathy for Lamarck's approach). A later French mineralogist and zoologist, Étienne Geoffroy Saint-Hilaire (1772–1844) developed the argument that, since all animals have similar physiological structures, they must at some point be related, and that the higher forms must (somehow) have arisen from the lower ones.

Hence, by the first part of the nineteenth century it is possible to see that a number of evolutionary ideas were in currency and that a degree of cross-pollination of these ideas was in process. It has, for instance, been noted by many that Lamarck's ideas were very close to those of Erasmus Darwin, whom Lamarck may well have read, given Dr. Erasmus's then fame. That fame may well have spread to Goethe in Germany, whose profile as a writer was similar to Dr. Erasmus in that his wide interests included science as well as his creative works (although he did not blend them into his literary *oeuvre* as did Dr. Erasmus or Lucretius). In his *Essay on the Metamorphosis of Plants* (1790), Goethe argued that plant life was in a state of flux, one species deriving from another, and that all life—plant, animal, and human—derived from a single source.

In a case of wheels within wheels, the works of both Lamarck and Erasmus Darwin were well known to an Edinburgh tutor of Charles's, Robert Grant. An admirer of Erasmus Darwin, Grant did not believe that animals were creatively designed, and he talked of sponges as the ancestors of higher animals, an idea similar to Lamarck's conviction that higher animals had evolved from simple worms. Charles is known to have had prolonged conversations with Grant, as indeed he did with his

Cambridge tutors.[38] Hence, it is likely that the discussions of Grant and Darwin ranged widely, to include Grant's then-heterodox views on life and evolution.

Dr. Erasmus Darwin's stature at the time both as naturalist and poet might come as a surprise to people of the twenty-first century, but it tells us much about his stature in the eighteenth century that William Paley's *Natural Theology* was written in part to defend the argument from design against Dr. Erasmus's claim that adaptation was a natural process resulting from the purposeful activities of living things (the doctrine commonly referred to as Lamarckism although, as noted above, Lamarck may well have derived the idea from Erasmus Darwin). Another clear indicator of his contemporary stature is the fact that Wordsworth and Coleridge targeted him in their Preface to the *Lyrical Ballads* (1798). That famous manifesto of the Romantic credo disavows the elegant but rather high-toned Augustan phraseology used by Dr. Erasmus in favor of the more natural language of common people. The acknowledgment, albeit negative, of the fame and pre-eminence of Dr. Erasmus by two celebrated Romantic poets makes it probable not only that Lamarck read him but also that a good deal of what Erasmus's grandson was to publish one day was derived from Erasmus too, as this section of one of the grandfather's poems would seem to indicate:

> Organic Life beneath the shoreless waves
> Was born and nurs'd in Ocean's pearly caves;
> First forms minute, unseen by spheric glass
> Move on the mud, or pierce the watery mass:
> These as successive generations bloom.
> New powers acquire, and larger limbs assume;
> Whence countless groups of vegetation spring,
> And breathing realms of fin, and feet, and wing.
> Thus the tall Oak, the giant of the wood,
> Which bears Britannia's thunders on the flood;
> The Whale, unmeasured monster of the main,
> The lordly lion, monarch of the plain,

The eagle soaring in the realms of air,
Whose eye undazzled drinks the polar glare,
Imperious man, who rules the bestial crowd,
Of language, reason, and reflection proud,
With brow erect who scorns this earthy sod,
And styles himself the image of his God;
Arose from rudiments of form and sense,
An embryon point, or microscopic ens![39]

In the context of Charles Darwin's precursors, mention should also be made of another work in the field written by an (at first) anonymous English-speaker which, although more or less wholly forgotten now, in its day fully lived up to the name of "Victorian Sensation" (the title of James Secord's voluminous study).[40] The author, Robert Chambers, fearing ecclesiastical opprobrium, very effectively kept his name concealed right up until his deathbed (although Charles Darwin correctly guessed his identity a few years after its publication). The book was entitled *Vestiges of the Natural History of Creation* (1844), and it provides an encyclopedic summation of some then-leading theories and discoveries in the fields of biology, cosmology, geology, and other specialist fields, written by a well-informed layman for a mainly middle-class readership. When Darwin read it, he was a little taken aback by the similarity of some of its ideas to his own, for which reason I shall give a brief summary to facilitate the comparison.

From the start Chambers fixes his colors to the mast, rejecting the Judeo-Christian narrative as put forth in the King James Bible. In its place he puts a strictly materialist explanation for the creation of life on earth. He writes that the whole of the firmament was initially a diffused mass of "nebulous matter" (intergalactic gases). Over an unknown period of time, the stars, galaxies, sun, and Earth formed. The absence of any traces of plants and animals from metamorphic rocks shows that during Earth's early history "excessive temperatures prevailed," which could not have supported life. Citing the work of the foremost geologists of the time, Lyell, Sir Roderick Murchison, and Darwin's Cambridge tutor, Adam Sedgwick, Chambers concludes that organic life began after the appearance of dry land, which emerged over the eons through a combination of sedimentation and an upward thrusting of rock by forces not yet properly understood.

With dry land there then emerged "a theatre for the existence of plants and animals." The simplest forms emerged first, followed by the more complex. He stated that the Almighty would not have brought forth each individual species through the exercise of "immediate exertion." The Earth and the whole solar system came not from a one-off divine creation but from natural laws which were still, Chambers is careful to point out, "the expressions of His will." He develops that thought by explaining that it would be a narrow view of the Deity, "characteristic of a humble class of intellects," to suppose Him acting in particular ways for particular occasions. Chambers instead lends his support to the doctrine of "secondary creation" via natural regularities that assume the function of divine mandates.

Chambers saw a linkage between the simplest and more complex beings in the Great Chain of Being (a Medieval concept rooted in the ancient idea of a *scala naturae*). To explain this connectedness among different animals, he proposed transitional forms between species. This ladder of organic life did not appear all at once, as many biblical creationists believed, but developed over eons of geologic time. He stresses the biological continuum linking beast and man by explaining that the human fetus shows a similarity to that of an ape but that these features are "suppressed" before the baby is born and the infant goes on to become a "true human creature."[41] In his peroration, anticipating the hostility that this latter contention would produce, he attempts to head off objections rooted in what he viewed as "ignorant prejudice" against lower animals by pointing out that all creatures great and small are a part of the Divine Conception: "Let us regard them in a proper spirit, as parts of the grand plan."[42]

The Victorian public was transfixed by *Vestiges* (as it became known in the absence of any identifiable author). Darwin's future rival, Alfred

Russel Wallace, reading it in the autumn of 1845, was electrified by its arguments. Chambers's theory of progressive development convinced him of the truth of organic evolution. And precisely because the myriad of new facts and conjectures advanced by Chambers did not actually explain how the various animals and plants had assumed their distinctive character, Wallace was inspired to work at the "how" of the origin of species for himself. Darwin, distancing himself from the negative reviews of *Vestiges* penned by orthodox Christian readers such as Cambridge geology professor Adam Sedgwick, and seeing all too clearly the similarity of the Chambers argument to his own, diplomatically gave measured praise to a volume which, as he put it, "if it does no other good, spreads the taste for natural science."[43]

The Elusive First Step

AMONG THE general public *Vestiges* became something of a *succès de scandale* with its scripturally unattested notions of creation-without-a-creator and of nature operating largely independently to produce its plethora of life forms. It has been plausibly suggested that the furor caused by *Vestiges* in the public mind may even have deterred Darwin from including in his own work any extended consideration of the spontaneous creation of life, even though he was drawn to the idea; yet the absence of any such explanation in the work could easily be seen as a deficiency in his attempt to provide a comprehensive naturalistic explanation for the origin of all organic life.

Perhaps cognizant of the deficiency, Darwin conjectured privately about the spontaneous origin of the first life. He did so in a letter sent to botanist Joseph Hooker in 1871. There Darwin imaginatively outlined a scenario for the purely naturalistic origin of the first living organism, and did not seem to see the step as an insurmountable obstacle to his effort to provide a purely naturalistic explanation for the development of life, even though he could not appeal to the wonder-working mechanism of chance variation and natural selection until a self-reproducing biological machine had first arisen. (One cannot, after all, have anything new for nature to select from, or invoke competition among offspring, until there exist offspring with variations.)

In Darwin's day, a single cell was thought of as something fairly simple, so its chance origin seemed quite plausible to many, even after renowned French scientist Louis Pasteur put to rest the notion that spontaneous generation of life from non-life was a common thing. Darwin, acknowledging the state of knowledge in the wake of Pasteur's findings, conjectured in his letter to Hooker that while we do not find life springing from non-life today, it might have happened on the early earth, given that the first such organism would have no other life forms gunning for it:

> It is often said that all the conditions for the first production of a living organism are now present, which could ever have been present.—But if (& oh what a big if) we could conceive in some warm little pond with all sorts and ammonia and phosphoric salts,—light, heat, electricity etc., present, that a protein compound was chemically formed, ready to undergo still more complex changes, at the present day such matter would be instantly devoured, or absorbed, which would not have been the case before living creatures were formed.[44]

Problem, he hoped, solved—just so.

Today, too, there are evolutionary expositors who make light of the origin-of-life challenge. Richard Dawkins, perhaps Darwin's best known modern expositor, confidently describes the first emergence of life on earth as a "gradual, step-by-step transformation from simple beginnings, from primordial entities sufficiently simple to have come into existence by chance."[45] But the situation today is decidedly different from what it was in Darwin's time, however assured Dawkins may sound. We might, therefore, do well to pause over the truth status and indeed even the logic of Dawkins's notion of "entities sufficiently simple to have come into existence by chance," and establish whether such a notion can be supported by experimental evidence—especially since recent advances in molecular biology show that the humblest bacterium contains more genetic information than the instruction manual for NASA space probes. The very notion of a simple biological entity has become deeply problematical with our increasing knowledge of the molecular world in the last half century, and one might therefore wish to query whether such a thing can exist in nature.

Happily, some decades after Darwin, advances in laboratory technology made it possible to begin testing the claim. The best known experiment to investigate the possibility of life originating spontaneously on the early Earth was carried out by Stanley Miller and Harold Urey of the University of Chicago in 1953. On the face of it, it might appear incongruous that modern-day scientists would touch this subject with a barge pole. Up until the middle of the nineteenth century, to be sure, a form of pseudo-scientific folk-belief was doing the rounds according to which rotten material and even soiled linen was supposed to be able to induce the formation of small life forms. But as noted above, renowned French scientist Louis Pasteur finally and decisively put to rest the theory of spontaneous generation: only life can produce life, he demonstrated. Strangely, though, the outmoded faith in spontaneous generation did not die out completely, and both the Russian biologist Alexander Oparin and the British scientist John Haldane revived the idea in the 1920s. The somewhat questionable logic behind the 1953 Miller-Urey experiment—which from the perspective of posterity appears to have been a rather desperate venture—has been described as a trial to find out if life-from-nonlife, although far from usual, perhaps "did belong to the realm of the unusual and long ago,"[46] and whether state-of-the-art 1950s know-how could succeed in discovering an answer where predecessors had failed.

Miller and Urey theorized that if the conditions prevailing on the primeval Earth were reproduced in laboratory conditions, such conditions might prove conducive to a chemical synthesis of living material. In accordance with the best scientific information at the time, they filled their laboratory receptacle with methane, hydrogen, ammonia, and water—all of which were thought to have been constituents of that early terrestrial atmosphere whose conditions the pair were attempting to simulate. At this point, an electric spark was passed through the chemical mixture to simulate what scientists term "an energetic event," that is, the kind of energy which could have come from thunderstorms on the primeval Earth. The resulting liquid turned out on analysis to contain amino acids which, though not living molecules themselves, are the building blocks of proteins, essential to the construction of life.

In 1953 there were high expectations that the next step from amino acids might lead to the first replicating organisms. The media of the time certainly hoped so, with *TIME* magazine reporting of the two experimenters: "What they have done is to prove that complex organic compounds found in living matter can be formed.... If their apparatus had been as big as the ocean, and if it had worked for a million years instead of one week, it might have created something like the first living molecule."[47]

Astronomer Carl Sagan adjudged the experiment an important first step in the direction of the actual creation of life, declaring that "the Miller-Urey experiment is now recognized as the most significant step in convincing many scientists that life is likely to be abundant in the cosmos."[48] The experiment was kept at the forefront of people's attention by continuing reportage in the press, and found its way into school and university biology textbooks and museum displays. Thus was the impression fostered that an energy source could indeed initiate a reaction leading to the formation of life's building blocks.[49]

Of Myths and Monsters

THE NOTION that the building blocks of life were easily gotten may well have seemed intuitively "right" to the many journalists and members of the public acquainted with Mary Shelley's *Frankenstein* (1818) or with the classic 1931 film of the same name—all the more so since the imaginative genesis of Mary Shelley's science fiction appears to have had a substantial basis in science fact. In a recent study, *Raising the Dead: The Men who Created Frankenstein*, Andy Dougan claimed to have found an historical prototype for Baron Frankenstein. He noted that poet Percy

Bysshe Shelley, in his preface to his wife's novel, makes reference to "Dr [Erasmus] Darwin and the physiological writers of Germany" whose work, Shelley stated, suggested that the story that followed was "not of impossible occurrence."[50]

Who Shelley had in mind when referring to these German writers is not entirely clear. The names of Alexander Humboldt and Johann Wilhelm Ritter have been mooted, but Dougan points to two other candidates, the first being a professor of surgery and Royal Prussian physician from 1817–1829, Karl August Weinhold, whose *Experiments on Life and Its Primary Forces through the Use of Experimental Physiology* had appeared in 1817. In his publication, Weinhold describes a number of frankly bizarre experiments on dead animals which, upon receiving electrical shocks, "revived" in the limited sense that the corpses exhibited involuntary spasms. He also contended that electricity could revive brain function and restore the dead to life, although his experiments were conducted behind closed doors in his university laboratory and no proof was offered of his claim.

In my view, however, it seems equally likely that Dougan's other mooted candidate, Percy Bysshe Shelley himself, might have been the prototype of the restless over-reacher. Shelley was certainly an important *éminence grise* behind his wife's creative endeavors. He is known to have consulted many treatises on electricity and galvanism. Interested in Paracelsus, the sixteenth-century alchemist and physician, and also in Sir Humphry Davy's theories on the conversion of dead matter to living, the poet himself carried out experiments with electricity (to the extent of electrocuting *himself*),[51] which he understood to be the animating force of life.

His wife's book, which Janet Browne records as having been inspired in part by Percy Shelley's talk of Erasmus Darwin's preserving a piece of vorticella "in a glass case till by some extraordinary means it began to move with voluntary motion,"[52] was subtitled *The Modern Prometheus*, a description suggested by the hubristic figure of Ovid's *Metamorphoses* who stole "particles of heavenly fire" (probably meaning lightning) from the abode of the gods.

When Ovid took over the myth associated with Aeschylus and Hesiod in the Greek tradition, he (or possibly unknown Roman predecessors) developed it to make of his Prometheus a figure who creates and manipulates men into life: a *plasticator*. The particles of heavenly fire were the means by which he quickened his clay images into life, a conception of (re)animation which occurs in an only slightly different form in *Frankenstein*. Also of note in this context, Mary's husband would go on to compose a lengthy poem featuring Prometheus, a verse drama that placed the mythical figure in a decidedly more positive light. (More on this in Chapter 6.)

Experiments with galvanism were not uncommon in the first three decades of the nineteenth century. In the same year *Frankenstein* appeared, Adam Ure in Glasgow set out to "reanimate" the executed criminal Matthew Clydesdale (causing convulsions but little else). Interest in such experiments waned towards the middle of the nineteenth century, but surprisingly, just six years before the 1952 Miller/Urey experiment, Robert Cornish of the University of California had everything in place in a university laboratory to attempt to reanimate by electric shock the corpse of a recent death row inmate, one Thomas McMonigle—a procedure he would have carried out had not the University authorities sensibly stepped in to halt proceedings.

An Increasingly Hopeless Monster

GIVEN THE presence deep in even educated persons' collective imagination (which the Germans term "versunkenes Kulturgut") of a tradition of re-animation, the media interest in the Miller/Urey experiment is unsurprising. However, the complete chemical pathway devoutly hoped for by many in the wake of the Miller-Urey experiment was not to materialize. In fact, the unlikelihood of such a materialization was underscored in the very same year that the Miller-Urey experiment took place, when Francis Crick and James Watson succeeded in identifying the famous double helix of DNA. Their discovery revealed, among other things, that even if amino acids could somehow be induced to form proteins, there was more to the story. Life also depends on nucleic acids, one of which is deoxyribonucleic acid or DNA, where the vital information needed to replicate and operate any given organism is encoded.

Proteins and DNA must be able to work together. DNA is both highly complex and highly specific, to the extent that just small differences in its letter sequences can make the difference between a living, thriving animal and a stillborn. Proteins are indispensable, but they do not have the capacity to store and transmit information for their own construction. DNA, on the other hand, can store information but cannot manufacture anything or duplicate itself. It's a chicken-and-egg situation, so much so that Francis Crick was once moved to comment that the beginnings of life seemed impossible, barring a miracle, since "so many are the conditions which would have to be satisfied to get it going."[53]

Finally, it had to be conceded that life was unlikely to form at random from the so-called "prebiotic" substrate on which scientists had previously pinned so much hope. (To this day, biochemists remain ignorant of the modalities of a jump from amino acids to proteins, and the origin of nucleic acids is similarly shrouded in darkness.) To complicate things even further, it is now widely disputed whether the early atmosphere of the Earth postulated by Miller and Urey would have been such as they assumed, and so it may not have supported the formation of the organic compounds they identified. Hence the problem appears now to extend to include the origin of the basic building blocks themselves.

The hope that life may be somehow "dormant" in chemicals, waiting to be unlocked when the correct combination of chemicals clicks into place, as it were, has clearly suffered a signal reverse. The large claim that life could arise from a reaction within any chemical substrate possessing the requisite prebiotic properties has, in addition, failed to be confirmed by recent space exploration. Had the twentieth-century Viking space mission been successful in finding evidence of even rudimentary life on the Martian surface, it might have been taken as confirmation by analogy that terrestrial life had emerged from a comparable chemical matrix. However, the space searches found no incontestable evidence, and the failure to find evidence of "exobiologies" based on exotic chemistries means that there is less confidence now than there was in the 1980s that some autonomous "cosmic imperative" might prompt the production of life wherever the "right" geochemical conditions prevail. The essential question of how lifeless chemistry might be translated into living biology remains unresolved.

More recently it has been mooted that deep-sea thermal vents may have encouraged early life-forms, or that certain types of clay could have encouraged prebiotic chemicals to gather, but neither of these options has prompted much scientific support to date, and most experts have had to concede that the step from a barren, primordial world to one of life-producing chemistry remains imponderable.[54]

It is then curious that Dawkins can claim that "the results of these experiments have been exciting." In the following passage he seems to place inordinate faith in ideas which have yielded only negative results to date:

> Organic molecules, some of them of the same general type as are normally only found in living things, have spontaneously assembled themselves in these flasks. Neither DNA nor RNA has appeared, but the building blocks of these large molecules, called purines and pyrimidines, have. So have the building blocks of proteins, amino acids. The missing link for this class of theories is still the origin of replication. The building blocks haven't come together to form a self-replicating chain like RNA. Maybe one day they will.[55]

As noted by American philosopher Thomas Nagel, a non-theist, "I find the confidence among the scientific establishment that the whole scenario will yield to a purely chemical explanation hard to understand, except as a manifestation of an axiomatic commitment to reductive materialism."[56] In fact, in a 2020 paper entitled "We're Still Clueless about the Origin of Life," Rice University professor of synthetic organic chemistry James Tour advocates for a moratorium on origin-of-life research on the grounds that "its overexpressed assertions jeopardize trust in scientific claims in general."[57]

It Came from Outer Space

SPECIAL PLEADING-CUM-WISHFUL thinking only serves to reinforce the impression that the origin-of-life problem is likely to remain insoluble in scientific terms, an impression reinforced by ever more speculative explanations for the origin of life. One such alternative posits that the first life on planet Earth was seeded from outer space. This idea, officially dubbed "panspermia," was first put forward in 1903 by a Swedish scientist, Svante Arrhenius, who proposed that microbes ejected from unspecified planets harboring life traveled through space and alighted on Earth. In 1973 Francis Crick and Leslie Orgel concluded that the Arrhenius theory was unlikely, but then they put forward the even more imaginative theory of "directed panspermia," according to which an advanced civilization somewhere in the Milky Way seeded the Earth and other planets with microorganisms.

Latterly the panspermia idea has been most commonly associated with the late Sir Fred Hoyle and his younger colleague, Chandra Wickramasinghe. Their basic idea is that life became seeded on Earth from outer space, wafted in our direction by intergalactic forces, the equivalent of cosmic convection currents, or perhaps attached to, or embedded in, meteors that came to Earth. The idea was dubbed "Hoyle's Howler" by no few biologists,[58] for even in the unlikely event it were true (which even Hoyle himself later doubted), it does nothing to elucidate the mystery of the origin of the first living organism. It simply relocates it to outer space, a fact emphasized by Michael Denton:

> Nothing illustrates more clearly just how intractable a problem the origin of life has become than the fact that world authorities can seriously toy with the idea of panspermia.
>
> The failure to give a plausible evolutionary explanation for the origin of life casts a number of shadows over the whole field of evolutionary speculation.[59]

What is particularly instructive about the panspermia idea is the motive which gave rise to the conception in the first place. Those few old enough to remember Hoyle's avuncular but no-nonsense manner on BBC TV and radio programs in the 1960s and '70s will remember his famous analogy concerning the greater likelihood of a tornado sweeping through a junkyard and accidentally assembling a Boeing 747 than of human life forming spontaneously on Earth. Viewers may also have read the first two chapters of his book, *The Intelligent Universe: A New View of Creation and Evolution,* and chuckled at the bluff, North Country manner of his demolition of all things Darwinian in its first two chapters. "How," Hoyle asks, "has the Darwinian theory of evolution by natural selection managed, for upwards of a century, to fasten itself like a superstition on so-called enlightened opinion?" His response consists in another rhetorical question followed by his own response, which also takes on Darwin's "warm little pond" conjecture:

> So why do biologists indulge in unsubstantiated fantasies in order to deny what is so patently obvious, that the 200,000 amino acid chains, and hence life, did not appear by chance? The answer lies in a theory developed over a century ago which sought to explain the development of life as an inevitable product of the purely local natural processes. Its author, Charles Darwin, hesitated to challenge the Church's doctrine on the creation, and publicly at least did not trace the implications of his ideas back to their bearing on the origin of life. However, he privately suggested that life may have been produced in "some warm little pond," and to this day his followers have sought to explain the origin of terrestrial life in terms of a process of chemical evolution from the primordial soup. But, as we have seen, this simply does not fit the facts.[60]

Hoyle's words clue us in on how abiogenesis and subsequent evolutionary processes must necessarily go together like the proverbial horse and carriage, and also give a good indication as to why Hoyle should have been driven to develop the rather desperate theory of panspermia in succeeding chapters (which must have shocked his readers in 1983 for whom Hoyle's BBC persona was the very apotheosis of common sense). Hoyle likened the statistical possibility of life appearing spontaneously on earth to a blindfolded person randomly solving a Rubik's cube: "If our blindfolded subject were to make one random move every second, it would take him on average three hundred times the age of the earth, 1,350 billion years, to solve the cube." The spontaneous origin of life on Earth, according to Hoyle, faced odds at least that long, if not longer. Such difficulties, Hoyle ruefully remarked, meant that "the scientific facts throw out Darwin but leave William Paley, a figure of fun to the scientific community for more than a century, still in the tournament with a chance of being the ultimate winner."[61]

In retrospect we can see that there was more at stake in the Miller-Urey experiment than many may have realized at the time. Its implicit promise was that it would extend Darwin's narrative timeline back to the pre-organic formation of the first cell of life, and so establish the fundamental point of departure for the joint mechanism of random variation (among offspring) and natural selection to go to work. The failure of this and of later, similar experiments have removed an indispensable foundation stone, a *sine qua non*, for the operation of natural selection.

3. THE CHALLENGE OF INTELLIGENT DESIGN

> If the world's finest minds can unravel only with difficulty the deeper workings of nature, how could it be supposed that those workings are merely a mindless accident, a product of blind chance?
>
> —Theoretical physicist Paul Davies[1]

Paul Davies's admirably direct question cited above poses an uncompromising challenge to sundry received ideas in the evolution field. In its entirely proper deference to the niceties of academic debate, however, his question still contains an element of diplomatic understatement. For it is not just that nature's workings can be unraveled "only with difficulty"; in very many cases they cannot be unraveled at all. Unlike cosmologists who confess (often with cheerful and unprompted candor) to having little more than a clue about dark energy, dark matter, black holes, and other cosmic arcana, and that anyway only at most 4% of the visible universe is even available to be directly observed and studied by them, Darwin and many later biologists have tended to paper over the cracks of their own lack of knowledge. This lack of transparency has inevitably led to a number of scientific objections, many arising in the later decades of the nineteenth century, others in the more modern period.

The claim that Darwinism gives us the key to unlock what some of Darwin's scientific peers were calling "the mystery of mysteries"[2]—the origin of life's great diversity of forms—was most signally challenged in the immediate aftermath of the publication of his *Origin*. A brief reprise of older criticisms will, I hope, prove useful in the interest of historical contextualization, foreshadowing as they do several objections to Darwinism which have resurfaced within the last half century. These older criticisms are especially useful since they date to a period when Darwin had not yet been enthroned in popular consciousness as The Sage of Down House (his country home). These earlier reviewers were thus less minded to pull their punches.

Near-contemporary reviews of the *Origin* were in the main negative. Søren Løvtrup goes so far as to observe that Darwin's "theory was rejected by almost all who had the power to judge,"[3] and a curious aspect of the work's immediate reception is that two allies of Darwin, Thomas Huxley and Joseph Hooker, eminent biologists both, were loud in their support of Darwin but remained unpersuaded by his special theory of natural selection. As David Knight put it, many of the intellectuals who first responded to Darwin's work concluded that since "nobody had ever seen one species change into another... Darwin's was not a proper inductive generalisation; and no definite outcome could be deduced from his law and tested by experiment, as the existence of the planet Neptune had recently been predicted from the Newtonian theory of gravity."[4]

Huxley had a problem with the very purposelessness claimed for the selection process. How was it possible for a process knowing nothing of the final end of things to fulfill its winnowing role when the only test *is* that end? He also attempted to reintroduce what Darwin regarded as the "t" word (teleology) into the equation by suggesting that natural selection did have a purpose, which consisted in the well-being and progress of the group or species to which the individual belonged even if not necessarily of the individual itself.

Huxley, however, consented to becoming a pseudo-Darwinian (in Peter Bowler's term) out of philosophic necessity: he was a fervent materialist lending his support to any movement subserving his non-theist cause, and he viewed Darwin's work as "a veritable Whitworth gun" for exploding religious dogma.[5] For if one has a conception of nature and planet Earth as a closed, autonomous system unsupervised by a divinity or any other force, then *some* wholly naturalistic process of species development was an ideological and indeed logical necessity. Whether

Darwin had hit on the *right* natural solution might well have been a secondary consideration to Huxley, given his ideological agenda.

A second curious factor about the very early reception of the *Origin* was the reaction of the reader chosen by Darwin's publishing house, Whitwell Elwin. That reaction has since become the occasion of much mirth, since he advised rejection of the manuscript, recommending that the author should confine himself to pigeons: "all the rest should be abandoned." Elwin, respected editor of *The Quarterly Review*, found the work "a wild and foolish piece of imagination," vitiated by its own over-speculative tendencies: "At every page I was tantalized by the absence of proofs... It is to ask the jury for verdict without putting the witnesses in the box."[6]

Coming now to the mainstream reviewers, historian Janet Browne[7] notes that the leading philosophers John Herschel, William Whewell, and John Stuart Mill also thought the work massively conjectural, with few findings that could be claimed as proofs. (Darwin was particularly stung when it got back to him that Herschel had dubbed natural selection "the law of higgledy-piggledy."[8]) The novelist George Eliot was lukewarm, pointing out that the volume was "sadly lacking in illustrative facts."

Harvard professor Asa Gray and geologist Charles Lyell made the plea for some sign of teleology, rather than mere chance, in evolutionary theory, and were supported by St. George Mivart. Mivart reasoned that, just as there is a principle internal to an organism which determines its embryological development, so must there be a similar, *internal* principle to determine the evolution of a species as a whole. Mivart here seems to echo an originally Aristotelian idea of immanent teleology, a doctrine which became influential for European posterity in opposition to the randomness of the Epicurean and Lucretian philosophies, which fell into neglect in Europe until the eighteenth century, as observed above.

Then, too, there was the foremost anatomist of the first half of the nineteenth century, Richard Owen, who also believed that the primary evolutionary force was internal to the organism, rather than occurring by natural selection. Like the earlier French naturalist Cuvier, Owen had assigned humans (in contradistinction to apes) to the separate taxonomic group of *archencephala* (of superior brain): a rather crucial distinction.[9] In the same year Moritz Wagner cautioned that speciation *in situ* was inconceivable since any differences that emerged would become genetically absorbed and leveled out over succeeding breeding generations with the inevitable return to the *status quo ante*, that is, a reversion to the genetic mean. Any chance of lasting variation could therefore only be envisioned on the basis of genetic isolation taking the form of a migration away from the original animal or tribal grouping.[10]

Mivart was and remained Darwin's principal opponent, developing his Darwinian critique into a full-scale monograph a decade after the *Origin*, entitled *The Genesis of Species*. In that volume he itemized his objections, the chief of which included the claim that "natural selection is incompetent to account for the incipient structures."[11] Mivart also advanced the idea that specific differences could be developed suddenly instead of gradually, this being the idea of "saltations" or sudden spurts, against which Darwin set his face. There were also definite limits to the variability of species, he continued, in opposition to Darwin's idea of well-nigh limitless transmutations over time, including the cross-over to new species.

Mivart also pointed out that "certain fossil transitional forms are absent which might have been expected to be present."[12] This latter point represented a major stumbling block then, as it does now, to the acceptance of Darwin's theory, and it was based on good evidence which Darwin had access to. In the year before the publication of Darwin's *Origin*, Edward Hitchcock, in his volume *The Religion of Geology*, had found that the fossil record did not show a gradual development of life forms via intermediaries but rather a discontinuous start-and-stop process involving just those kinds of "saltations" that Darwin ruled out of account.[13] Hitchcock's conclusion that these discontinuities in the fossil record were an indication of repeated divine interventions was of course precisely the doctrine which Darwin was determined to oppose, even, it appears, at the cost of ignoring important evidence if it undermined his own position.

Separate mention must also be made of that critic whom Darwin respected but feared almost as much as the pugnacious Mivart: the Scots engineering professor Fleeming Jenkin, a man of limitless ingenuity who had obtained more than a score of patents for his inventions. In his 1867 review of Darwin's work[14] Jenkin did not doubt the possibility of minor changes like the ones Darwin had studied in the avian population of the Galapagos Islands. Slight improvements "making a hare a better hare or a weasel a better weasel" were eminently possible: natural selection in that limited sense was uncontroversial, he conceded; but this did not "imply an admission that it can create new organs, and so originate species." There was a limit to the amount of variation that could be anticipated, and the important factor of interspecific (hybrid) sterility would prevent the possibility of certain evolutions. As to Darwin's invocation of time itself as the factor which could account for the snail's-pace progress of speciation over long ages, Jenkin replied with refreshing briskness, "Why should we concede that a simple extension of time will reverse the rule?"

Towards the end of the century, Darwinian theories were rejected by the eminent scientist who introduced Mendelian genetics to British science, William Bateson, who concluded, "The transformation of masses of population by imperceptible steps guided by selection is, as most of us now see, so inapplicable to the facts… that we can only marvel both at the want of penetration displayed by the advocates of such a proposition and at the forensic skill by which it was made to appear acceptable even for a time."[15]

Darwinism in the Twentieth Century

In Vernon L. Kellogg's exhaustive survey of biological research to 1907, he concluded that Darwinism was "cast down" as a credible theory,[16] and in the early twentieth century there was a real possibility that it could have been eclipsed altogether had it not been rescued, as it were, by attempts to update it by synthesizing it with the emerging science of genetics—which Darwin knew nothing of despite a partial overlap of the lives of Darwin and Gregor Mendel, the Moravian monk who established the modern understanding of genetics.

On this view, the first three decades of the twentieth century were a low point in Darwin's reputation. While there is a case to be made against the idea of a "Darwinian eclipse," and against the idea that the neo-Darwinian synthesis rescued it from oblivion,[17] clearly there existed at the turn of the twentieth century a place at the commanding heights of our culture for challenging Darwin. In 1907, the same year in which Kellogg predicted the demise of Darwinism, an anti-Darwinian theory of biological origins was advanced by French philosopher Henri Bergson in his work *Creative Evolution*.[18] The level of that theory's acceptance is indicated by the fact its author was awarded the Nobel Prize. Bergson's theory postulated that all life results, not from mechanistic forces as Darwinism taught, but from a vital impulse that caused evolution, the *élan vital*, a non-material force guiding evolution in specific directions. The *élan vital*, Bergson explained, is a basic force like gravity or electromagnetism and, like them and many other physical phenomena, its origin cannot be explained.

Bergson also concluded that, due to what later writers would call irreducible complexity, at every stage of any given animal's evolution, all the parts of an animal and of its complex organs must have varied contemporaneously so that effective functioning was preserved (the idea first advanced by Cuvier). Thus it was implausible to suppose, as did Darwin, that such variations could have been constructed one tiny random step at a time. Bergson's influence on French science continued into the latter part of the century when Pierre-Paul Grassé's *L'Evolution du Vivant* (1973) came out in favor of a more directed form of evolution than that allowed for in the Anglo-Saxon and German traditions. Its author frankly asserted that "present-day ultra-Darwinism, which is so sure of itself, impresses incompletely informed biologists, misleads them and inspires fallacious interpretations."[19] He insisted that the ultimate mystery of evolution remained unsolved, and in the book's final sentence conjectured, "Perhaps in this area biology can go no farther: the rest is metaphysics."[20]

Thus it is that a work from the next decade, Australian geneticist Michael Denton's *Evolution: A Theory in Crisis* (1985), with its detailed and firmly argued critique of Darwinism, did not come entirely out of the blue. Indeed, he had others besides Grassé for recent company. In 1960s and 1970s Nobel Prize winner Sir Peter Medawar expressed misgivings about reductionist approaches to complex biological problems.[21] The novelist and scholar Arthur Koestler, who very effectively straddled the "two cultures" of sciences and humanities described by the scientist-novelist C. P. Snow in 1959, argued in the 1960s that natural selection was far too simplistic an explanation to account for many of nature's complexities. For him the extreme unpredictability of the posited natural selection process amounted to little more than a game of blind man's bluff. Given the number of "interlocking parts" which would have to come together in a perfectly synchronized way, he concluded that "the doctrine of their coming together due to blind coincidence is an affront not only to common sense but to the basic principles of scientific explanation."[22]

Evolution and the Humanities

Scholars from within the humanities proper objected to "neuro-evolutionary thought creating new biology-based disciplines encroaching on the intellectual territory of the humanities."[23] And David Holbrook, in his volume *Evolution and the Humanities* (1987), devoted one whole chapter specifically to "Rescuing the Humanities from Darwinism."[24] Holbrook, an English literature academic, held that this "toxic and essentially nihilistic metaphysic," which reduces life to meaninglessness, was not the best basis for reading, viewing, or listening to the crowning cultural achievements of mankind.

Even among evolutionary biologists unwilling to break wholly with Darwinism, there were prominent members of that community who pushed back against the scientism that many of Darwin's follow-ers seemed eager to aid and abet. World evolution authority Stephen J. Gould did not see his scientific specialty as the be-all and end-all. As Kim Sterelny noted, "Gould does not think science is complete. The humanities, history and even religion offer insight into the realm of value—of how we ought to live—independent of any possible scientific discovery."[25]

Such an outlook calls into question what Raymond Tallis has termed "full-on biologism," a commitment to explaining social behavior strictly by recourse to biological methods and principles, including the social behavior of humans. Others go further than Gould , balking not only at "full-on biologism," but more specifically at Darwin's assumption in his *Descent of Man* that humanity's complex linguistic capacities had evolved from some earlier form of apelike communication. Victorian readers did so from the start, as Alvar Ellegård found in his study of the British press in the later Victorian era.[26] The bare physiology of ape and human vocal tracts is very different, the human variant longer and differently configured[27] to facilitate those sequences of extended vocal-izations which we call language. The ape, by contrast, is physiologically constrained to be able to produce only a very limited range of sounds.

And even if it is perfectly possible to speculate that the esophageal physiology of the ape may have evolved in the direction of its human equivalent, this does not explain how the rapid mental processing on which articulate speech depends kept pace with that process. Synchro-nization of those two processes would of course point not to random evolution but to coordination and therefore design. How could the fa-cility of speech, which depends on the interdependent agency of brain, mouth, lungs, and tongue have developed by any process of natural selec-tion, which would have needed to ponderously reconfigure a vast suite of genetic changes in the genome and a corresponding set of neuronal changes in the brain?

Opposition to the Darwinian view came from two legendary lin-guistics specialists: Professor Friedrich Max Müller, who in the nineteenth century was a world expert at Oxford in the then very much "trending" field of Indo-European philology; and the American linguistic science expert Noam Chomsky, famed both in the linguistics sphere and for his activism in the political arena from the middle of the twentieth century onwards. In a series of lectures at the Royal Institution in 1860, Müller, although he expressed sympathy for some of Darwin's ideas, claimed that the language of primitive humans could not have developed from animal sounds, as Darwin supposed. He put forward the idea that words were related to mental concepts in a non-onomatopoeic way. It was, he claimed, impossible for language to arise from the vocalizations of animals, because animals manifestly had not developed any understanding of concepts.

Noam Chomsky argued the case for an inbuilt universal grammar which he thought to be embedded in the neuronal circuitry of the human brain.[28] He was convinced that language competency was largely innate, not something that had to be acquired after birth. Language was fixed in the form of inbuilt specified rules and a child will adapt appropriately to the relevant linguistic cues whether the child must speak English, or Chinese, or any other tongue. This is why we all pick up our first language with such ease, because there is a form of (as yet unidentified) language "organ." Finally, he concluded, the human language facility was not an adaptation but a "mystery" (Chomskyan code for a puzzle unlikely ever to be solved). Chomsky's contra-Darwinian position was encapsulated in his statement that people were welcome to say that language evolved if they wished, "so long as they realize that there is no substance for this assertion, that it amounts to nothing more than a belief."[29]

A Duel with the Dualists

MÜLLER AND Chomsky were not the only "celebrity" dissenters. The first, astoundingly, had been none other than the co-discoverer of the natural selection theory, Alfred Russel Wallace himself, whose later defection from the purer Darwinian faith was to blindside Darwin. In his earlier writings, Wallace's description of natural selection stays in lockstep with his peer's, without the slightest interpolation of anything smacking of metaphysics. In fact, for many years Wallace was thought to be an even stronger advocate for natural selection than Darwin himself, frequently arguing that all the rich complexity of life must have evolved naturally without benefit of intelligent guidance. However, he parted company with Darwin eventually on the subject of the human mind, citing his inability to comprehend how unconscious processes could produce consciousness. He came to think that human self-awareness was of such a high order of sophistication, and so unlike the things that had arisen by natural causes, that natural selection was powerless to account for it. He could not bring himself to believe "that the mere addition of one, two, or a thousand other material elements to form a more complex molecule, could in any way tend to produce a self-conscious existence."[30]

Not surprisingly, the journal *Nature* upbraided Wallace for his dualism, concluding scathingly, "to say that our brains were made by God, and our lungs by natural selection, is really to exclude the Creator from half His creation, and natural science from half of nature."[31] In a similar vein, Darwin roundly arraigned Wallace for writing "like a metamorphosed (in retrograde direction) naturalist, and you the author of the best paper that ever appeared in the *Anthropological Review!*"[32] Darwin was particularly scandalized to realize that Wallace had been taken in by the contemporary craze for séances,[33] something he and many others would find as inexplicable as Arthur Conan Doyle's being taken in by fake photographs of the so-called Cottingley Fairies a few decades later.

What Darwin construed as the infirmity of an otherwise noble mind (although it must be said Wallace's mind remained sharp and "research active" even into his late eighties) explained at a stroke for him the genesis of his colleague's defection. For in the course of communicating with deceased persons one was (ostensibly) communicating with disembodied entities—which are of course by definition non-corporeal. Later scholars have suggested just this causal chain, but there is a timeline problem. As Michael Flannery notes, Wallace first went public with his view that evolution must have a teleological component in 1864, at which time by his own account he was a skeptic of séances. And he didn't attend his first séance until the following year.[34] Be that as it may, Wallace became convinced that mind and body really were discrete entities, and it is no wonder that Darwin was dismayed by his colleague's defection. For if human language is thought to fall outside the evolutionary scheme, so does man himself—he clearly can't be half in and half out, his body brought about by natural selection and his brain fashioned by God! To accept Wallace's contention, then, would seem to sound the death-knell for Darwin's theory.

Further sharp criticism of Darwinian tenets has arisen in respect to the brain, the acknowledged jewel-in-the-crown of *homo sapiens*, a subject that inevitably brings in its train the intangible, non-material phenomena of consciousness, thought, and the subjective self. In the course of the last half century, neuroscience has shown that the human brain is something of such awe-inducing complexity that no hyperbole can do it justice. The hundred billion (yes, billion) neurons in our brains are all connected to other neurons by small fibers (dendrites) to allow instantaneous communication. There are more than a quadrillion electrical connections or synapses which make it the most complex piece of "machinery" known to mankind, though the term "machinery" fails utterly to capture its sophistication. Even describing it as a sort of wetware mainframe computer is a vastly belittling comparison. I defy anyone not to rub his or her eyes in wonder when reading Denton's description, reproduced here:

> Attempting to visualize a billion neurons, each a tiny nanoscale navigator, preprogrammed with a unique set of maps and the ability to match each map, at a defined and preprogrammed time, with the unique configuration of a series of unique sites in the ever-changing terrain of the developing brain, all homing in, unerring, toward their target, brings us indeed to the very edge of an "infinity" of adaptive complexity. The unimaginable immensity of "atomic maps," "molecular charts," "nanotimepieces" and other nanodevices used by this eerie infinity of nanorobots which navigate the ocean of the developing human brain, building as far as we can tell the only machine in the cosmos that has

genuine understanding, is far greater than that of all the maps, charts and devices used by all the mariners who ever navigated the oceans of earth, far more than even all the stars in our galaxy, more than all the days since the birth of the earth.[35]

Even that fine description evokes only the physiological mechanisms of the brain. Notwithstanding science's growing knowledge of the brain's physiological structures, it cannot purport to explain how the firing of literally billions of neurons translates into thought and emotion. Subjective experience, *mind*, appears to occupy some as yet invisible and indeed unvisualizable order of reality inexplicable in terms of material properties. In philosophical parlance, this introduces the issue of "dualism," the question of whether there can be a physical world (one knowable by empirical observation) plus an additional world of consciousness and self-knowing which resists such probing. A dualistic view was espoused by philosopher Sir Karl Popper in the latter part of the twentieth century, whereas orthodox Darwinism tends to regard humankind's self-consciousness as a mere epiphenomenon, meaning an accidental outcome of the mechanical workings of the brain.

For Darwinian dissenters it simply lacks logical coherence to suppose that *sentience*, baffling to the best scientists and thinkers to this day, could have evolved template-less from any purely material matrix. How could the interplay of impersonal forces have all unwittingly been instrumental in the creation of persons? It is not even possible to imagine a theoretical pathway leading to how consciousness could have come about by natural selection—which is one reason why leading scientists in the field, such as Susan Blackmore, have found it threatening to the Darwinian paradigm.[36]

Scientists often adopt hushed and almost embarrassed tones in referring to consciousness, since it represents a major challenge to the materialist schema into which all facets of human life "should" be able to be fitted. They are loath to acknowledge that the problem rests with the Procrustean, one-size-fits-all schema they support. Not surprisingly then, for Denton, the large improbability of intelligent life being formed by forces bereft of all cognitive capacity provides nothing less than a "formal disproof" of the whole Darwinian dogma, dubbed by him "the great cosmogenic myth"[37] of the modern era.

Uncooperative Evidence

DARWIN'S MAIN hypothesis, concerning the idea of a biological continuum with its ascent-of-man narrative from ape to *homo sapiens*, has been cast into considerable doubt on a whole host of fronts. Empirically, the idea of a transformation from one species to another appears problematical in view of the practical experience of animal husbandry, where even selective breeding has proved unsuccessful in bringing about fundamentally new species.

Darwin must also of course have known that sentient beings are not built up of discrete parts on the modular basis of prefabricated sheds. Cuvier had long since established what he termed the Law of Correlation, which stated that no part of a body could change without the whole changing, since all parts of any given body must be in perfect relation to its other parts. For a species to change dramatically there would have to be co-adaptive changes since, in the words of the old ditty, "The thigh bone is connected to the hip bone," and so on and so forth. The simultaneous co-adaptation of numberless body parts, including internal organ modifications and information storing systems to make all the reconfiguring function properly, is not easy to imagine, much less to actually engineer.

All these factors argue against the notion of major biological innovations via blind evolutionary processes.

If one takes this evidence as indicative rather than illusory, then here too Wallace was ahead of the game. His first break with materialistic evolution, recall, came when he espoused an awkward dualism in which the brain was created, but the rest of our mortal frames came about by natural selection. Late in life Wallace resolved this conceptual anomaly by espousing a full-on form of intelligently guided evolution, arguing that numerous innovations in the history of life were beyond the reach of random variation and natural selection, and their origin could only be adequately explained by reference to a teleological process. Indeed, he became convinced by his study of the evidence that not only humanity but the entire cosmos had been intelligently crafted.[38]

Darwin, in contrast, held out for an evolutionary process untainted by teleology. He sought to support his grand ontological step-change postulate from ape-like ancestor to humans by reference to the idea of "deep time." That is, given millions of years, all sorts of improbable and seemingly impossible things can happen, right?

We observed above, in the context of the early reception of the *Origin*, how the Scots professor Fleeming Jenkin discounted this proposed solution. The point Jenkin made in 1867 was reinforced by Hoyle in 1983:

> A generation or more ago a profound disservice was done to popular thought by the notion that a horde of monkeys thumping away on type-writers could eventually arrive at the plays of Shakespeare. The idea is wrong, so wrong that one has to wonder how it came to be broadcast so widely. The answer I think is that scientists wanted to believe that anything at all, even the origin of life, could happen by chance, if only chance operated on a big enough time scale.[39]

In his book *Not by Chance*, Lee Spetner makes a similar argument. "They think the earth's age is long enough for anything to have happened," he writes. "When one deals with events having small probabilities and many trials, one should multiply the two numbers to determine the probability. One should not just stand gaping at the long time available for trials, ignore the small probability, and conclude that anything can happen in such a long time. One has to calculate."[40]

Recourse to the lapse of millions of years to explain the evolutionary origin of humans from ape-like ancestors places the idea beyond the reach of any directly empirical test, and so the idea badly needs at least indirect evidential backup. Unfortunately for Darwin there is a dearth of fossil evidence to establish the claimed evolutionary "missing links," a situation not likely to improve in future time, to judge by the persistent dearth of credible missing-link hominid fossils found after his day, which have tended to be damaged and/or scattered. The ideal of finding a neat sequence of fossils layered over time to permit minute observation of continuities and discontinuities seems unachievable. The unfortunate result is that deciphering the course of human evolution from the fossil record has been likened to trying to work out the plot of *War and Peace* from about a dozen pages torn *at random* from Tolstoy's mammoth novel.[41] This has meant in practice that preconceived ideas have often been given license to play as much of a role as dispassionate observation in many fossil analyses.

What if we broaden our fossil search to all promising missing-link fossils, not just hominid ones? Even with this allowance, the number of stop-the-presses missing-link fossils that have been discovered are relatively few in number when set against the billions of transitional forms implied by Darwin's theory. The discovered fossils that have raised a buzz include the *Archaeopteryx,* thought to be a hybrid, half-bird, half-reptile; the *Eohippus,* a dog-sized ancestor of the horse thought to have evolved in size over 50 million years, and the well-known Neanderthal man and Cro-Magnon man remains. There have also been a number of forgeries, the most famous of which was "Piltdown man," a skull with a man-like brain and an ape-like jaw, which was for many years taken to be a vindication of Darwin's claim that there existed intermediaries between ape and man. This idea persisted from the time of its purported discovery in 1912 to the time it was revealed to be a hoax in 1953. A victimless crime it may have been in the financial sense, but it led many scholars down a bogus track for four decades.

Whether the Piltdown affair was a hoax of the April Fools variety (like the notorious "surgeon's photograph" of the Loch Ness Monster in the 1930s), or whether it was meant as an illicit remedy for what Darwin termed "the extreme imperfection of the fossil record," must remain moot. The persons associated with the excavations at Piltdown in southern England in 1912 were pillars of social responsibility: Charles Dawson, a solicitor, Arthur Keith, a respected anatomist,

Arthur Woodward, Keeper of Geology at the British Museum of Natural History, and Pierre Teilhard de Chardin, priest, amateur naturalist, and distinguished proponent of the theory of theistic evolution. It also seems unlikely that ancillary workmen helping in the excavation could have been responsible for a fabrication requiring at least a degree of paleontological sophistication.

Indeed, there is no definitive "smoking gun" suggesting that any of those upstanding persons were the perpetrators, and M. Bowden's identification of the Piltdown hoaxer with Teilhard de Chardin must finally remain unprovable.[42] Nevertheless, the affair surrounding this bizarre middle-class crime remains suspect, and as Gertrude Himmelfarb observed, "the zeal with which eminent scientists defended it, the facility with which those who did not welcome it managed to accommodate themselves to it, and the way in which the most respected scientific techniques were soberly and painstakingly applied to it, with the apparent result of confirming both the genuineness of the fossils and the truth of evolution, are at the very least suspicious."[43]

Curiously, Darwin himself acknowledged and indeed drew attention to the lack of fossil evidence—he even, as he put it, "had difficulty imagining by what gradations many structures had been perfected," adding, "Why, if species have descended from other species by fine gradations, do we not everywhere see innumerable transitional forms? Why is not all nature in confusion, instead of the species being, as we see them, well defined?... as by this theory innumerable transitional forms must have existed, why do we not find them embedded in countless numbers in the crust of the Earth?"[44]

Yet, as Gertrude Himmelfarb (who did more than any other critic to unmask Darwin's rhetorical evasions) noted, Darwin's technique here and elsewhere was "to assume that by acknowledging the difficulty, he had somehow exorcized it,"[45] coming up with a *faux* confession aimed at propitiating critical dissent. Thereafter, misgivings are whisked away by rhetorical legerdemain *cum* disarming self-effacement, and he proceeds, in a famously circular argument, to blame the fossil record itself for not providing the evidence he desired (lacing this with the pious hope that future fossil finds would prove him right). His hope that the gaps would be remedied after his day has not, however, been fulfilled to date.

In fact, a striking feature of the fossil record is that most new kinds of organism have tended to appear unheralded, in the sense that they are not led up to by a sequence of imperceptibly changing forerunners, as Darwin believed should be the case. A century of searching has only confirmed this pattern, such that by 1972, so grave was this anomaly perceived to be that Stephen J. Gould and Niles Eldredge put forward a theory of "punctuated equilibria" to account for the gaps—meaning that, *pace* Darwin, there must have been large spurts or saltations in evolution, followed by protracted periods of stasis in which the process is assumed to have become dormant. Only on that understanding could the many punctuations (gaps) in the fossil record be made sense of.

The idea ran counter to Darwin's insistence on a very slow and gradual process of evolution. For Darwin, his theory depended absolutely on an almost imperceptible rate of evolutionary growth, convinced as he was that a faster rate would be prima facie evidence of divine intervention. Perhaps remembering the analogy of Pallas Athene emerging fully formed from the head of Zeus in Greek mythology, he was adamant that, like Zeus, only God could perform saltations. Hence, had Darwin still been living in the 1970s, he might have concluded that his grand quest to establish a purely naturalistic narrative to account for life on earth had been compromised, if not entirely subverted, by the Gould/ Eldredge corrective.

What major fossil evidence we *do* have is that some 540 million years ago there occurred a phenomenon sometimes called the biological Big Bang, officially titled the Cambrian Explosion, which yielded a sudden emergence on earth of about thirty new basic animal forms (phyla)—not merely thirty species, but thirty fundamentally distinct body plans. A human and pigeon belong to the same phylum. That's how capacious the taxonomic category of phylum is. And there were some thirty new phyla in the Cambrian. These included arthropods, modern representatives of which are insects and crabs; echinoderms, including starfish and sea urchins; and chordates, the latter of which eventually gave rise to mammals. No transitional intermediates are to be found for any of these, and, furthermore, there is no evidence to show that the first species within each of these phyla did not emerge fully formed. This is yet again opposed to the idea of slow evolution, and the point has been made that Darwin's favored evolutionary metaphor, of a great Tree of Life arising from a tiny acorn with diverging branches, had come dangerously close to being uprooted by the Cambrian Big Bang.

Against Type

HAD DARWIN restricted his observations to Galapagos finches, there is no doubt his findings would have been welcomed without demur. It was (as the press reader had pointed out) the extrapolation from his avian studies to the whole living universe which seemed overly conjectural—a saltation which leap-frogged over all normal canons of logic. His intellectual gamble was essentially that, if all organisms are capable of evolutionary change, they could all undergo *unlimited* change, crossing some of the fundamental barriers apparently erected by nature and heretofore deemed impassable. He was making the staggering claim that the species barrier claimed to be unbridgeable by the biological typologists and practical animal-breeders alike in fact could be crossed. Darwin was taking a sledgehammer to the conception of fixed types espoused by the foremost anatomist of Britain at the time, Richard Owen, who together with other typologists thought that any variation could only ever be slight because it was constrained by the boundaries of each separate type, or species.

Darwin was in all but name sponsoring the idea of phylogenetic revolution, rather than just simple evolution.

That revolution would appear, from a cursory look around, to stand everywhere ascendent in our day. And modern Darwinism is indeed the dominant paradigm. But the present state of affairs is, well, complicated. In the last half century there has emerged a non-evolutionary mode of biological classification called cladism (clade = type, from the Greek for branch). Cladism does not make the presumption that species were ancestral to each other. Instead, it classifies organisms according to type without regard to any evolutionary assumptions. For a time it was associated in Britain with the name of Colin Patterson of the British Museum in London, who in 1981 was reportedly traveling the conference circuit and embarrassing fellow conferees with the acidulous question, "Can you tell me one thing about evolution that is true—any one thing at all?"[46]

Something of a brouhaha ensued at the museum in the early 1980s when cladism was used as a means of classifying exhibits, and the backlash caused the museum directors to capitulate, resulting in enforced retreats by museum staff, and the restitution of some more "appropriate" signage for exhibits in line with Darwinian conceptions.

The incident may seem in retrospect to be of minor significance, yet it nevertheless remains a straw in the wind, showing how easily the subject of evolutionary theory can divide opinion and how all is not as stable and settled in the house of Darwin as the official line might suggest. A consideration of Darwin's second major volume, *The Descent of Man*, may provide some further clues to the reasons for that resistance.

Darwin's The Descent of Man

THE DESCENT of Man,[47] being a logical pendant to *The Origin of Species*, put humanity explicitly at the forefront of investigation, whereas in the earlier book the relationship of evolution to man had been merely implicit. The twelve years between the books gave Darwin the opportunity to take issue with some of the criticisms leveled at his evolutionary theory, such as the dispute in the 1860s about the origins of language involving the philologist Max Müller, adverted to above. In taking issue with Müller's contentions in *The Descent of Man*, Darwin tends to talk up the cognitive capacities of the animal world, while at the same time playing down the abilities of humans to a degree most would find questionable. He writes, for instance, of a human infant having approximately the same mental development as a dog: "As everyone knows, dogs understand many words and sentences. In this respect they are at the same stage of development as infants, between the ages ten and twelve months, who understand many words and short sentences, but cannot yet utter a single word."[48]

Darwin even traces the religious instinct in humankind to an inchoate stage in the dog, as when, after his master returns home after an absence, the dog's reawakened feelings of adoration are such that it "looks on his master as a god."[49] As Himmelfarb noted, as Darwin "reduced language to the grunts and growls of a dog, he now contrived to reduce religion to the lick of a dog's tongue and the wagging of his tail."[50] Darwin even goes so far as to claim that a dog's "religion" is purer than that of a human, since human religion has been morally corrupted by such aspects as not only trial by ordeal and human sacrifice (an uncontroversial contention) but also by celibacy and prohibitions on certain foods—which he terms "absurd religious beliefs." Here Darwin is playing his part as a Victorian rationalist who neither accepted any Christian expressions of faith differing from the Anglican norm nor showed any respect for diversity in other faith traditions.

Most, I think, would find the intellects of dog and infant to be incommensurable, but, as James Moore and Adrian Desmond point out, "Darwin tended to humanize nature even as he naturalized mankind."[51] Why this should have been the case is not known, although there are clues. In 1827 he had attended a Plinian Society meeting in Edinburgh University where one attendee attempted to prove that "the lower animals possess every faculty and propensity of the human mind."[52] After a visit to London Zoo in 1832, Darwin felt an instinctive affection for what he termed the innocent creatures he viewed there. We also know from his reports from the *Beagle* that he felt some alienation from his human peers after witnessing the horrors of slavery together with what he termed the base conduct of some of the Fuegian people. Man's inhumanity to man seems to have convinced him that mankind should not "boast of his proud pre-eminence"[53] since man was in moral terms no better than the higher species of ape from which he had evolved.

There is no evidence from Darwin's everyday contacts with people that any settled misanthropy lay behind his attitude. Himmelfarb put it down to a form of subconscious "professional deformation" of the zoologist: "The practice of seeking explanations in the lowest common denominator—morality in terms of instinct, human motives in terms of animal impulses, and civilized conduct in terms of primitive customs—was perhaps a professional failing."[54]

Himmelfarb was writing in the late 1950s. Those of us who experienced the arrival of so-called "sociobiology" in the 1970s might have a rather different opinion. That era, it will be recalled, saw the publication of E. O. Wilson's *Sociobiology* (1975) and Richard Dawkins's *The Selfish Gene* (1976).

In *The Selfish Gene*, Dawkins famously encouraged people to think of themselves as "robot vehicles" epiphenomenal to their genes, puppets manipulated by their genetic makeup. These ideas which, as Roger Kimball asserts, "appeal to people who combine cynicism with credulousness,"[55] were vigorously opposed at the time by philosophers Mary Midgley and Antony Flew, who uttered the inconvenient truth that the selfish-gene hypothesis was neither true nor even faintly sensible. Both pointed out that it was logically perverse to claim that genes could be "engaged, whether selfishly, or unselfishly, in any conscious or chosen pursuit of anything."[56]

Wilson, meanwhile, was widely criticized for licensing the kind of thinking which would validate racism, eugenics, and sexism, even Nazism. However valid that particular charge may be, some of the responsibility for the worst excesses of "social Darwinism," with all its discriminatory and misogynist ramifications, must, alas, be placed at the feet of Charles Darwin himself. Himmelfarb writes of Darwin's "failures of logic and crudities of imagination" and "painfully naïve forms of analysis and exposition,"[57] which are precisely the simplistic and reductionist tendencies associated with sociobiology (later rebranded for tactical reasons as "evolutionary psychology").

Such weaknesses are particularly in evidence in the way Darwin treats the subject of women, whom he views, essentially, as less perfectly evolved versions of men. According to the Darwinian narrative (which was rightly disputed even at the time by a number of professional male colleagues in their discussions with Darwin), men's superiority was to be accounted for by the trials and tribulations they had successfully survived in "winning" females. Ostensibly battle-hardened by such conquests, males supposedly acquired a skill set which proved advantageous to them when transferred to other areas of life as well.

Did Darwin seriously believe that ancient man's sexual victories were habitually marked by violence and cunning, his enemies bloody at his feet and his conquest dragged off by the hair? As philosopher David Stove and others have pointed out, no mere slugging match or primordial *bellum omnium contra omnes* could have contributed to the survival of any human society of even minimal complexity.

This issue had in fact already been addressed by the Russian scientist Peter Kropotkin in 1902, in his aptly titled work *Mutual Aid: A Factor of Evolution*. His observations of harsh Siberian peasant life had revealed few signs of ruthless competition between group members. On the contrary, in the challenging conditions of life in that region of Russia at the time there was a premium on group members helping each other in order to preserve group cohesion and survival.[58] Kropotkin expressed open dissent from Thomas Huxley's unsubstantiated fantasy of primitive life having been little better than a gladiatorial combat, and pointed out that Darwin in his *Origin* had provided no shred of evidence for this speculation. (Not surprisingly, perhaps, since the idea was lifted from Malthus.) Modern fieldwork studies have supported Kropotkin's finding that those animals and humans which acquire habits of mutual aid are in truth the fittest to survive.[59] Cooperation rather than competition must always have played the greater role, something which even E. O. Wilson was ready to concede in a late Damascene conversion.

When Darwin makes the attempt to explain the crucial point of *The Descent of Man*, humankind's supposed descent from ape-like ancestors, he speculates somewhat vaguely on the question of whence we as a species got our superior brains: "The mental powers of some earlier progenitor of man must have been more highly developed than in any existing ape, before even the most imperfect form of speech could have come into use; but we may confidently believe that the continued use and advancement of this power would have reacted on the mind itself, by enabling and encouraging it to carry on long trains of thought."[60]

The passage has the disconcerting tone of a just-so story. How, one might legitimately ask, did one ape "happen" to get its superior cognitive capacities? What was the *vera causa* of its braininess? And how did this cognitive superiority trigger correlated changes in the brain? In the light of present-day scientific advances these seem like shallow assertions, inadequate to account for what we know about those labyrinthine co-adaptive changes necessary for the process he describes to function effectively.

On another point, this passage and many others like it would be a gift to linguistic specialists in discourse analysis or to those whose specialty is in the deconstruction of advertising propaganda. His reiteration here and elsewhere of the phrase "we may confidently believe" veils the tenuous truth-value of what he proposes, which is finally little better than a guess. This mode of assertion is uncomfortably reminiscent of the wearisomely repeated phrase of ex-PR-man turned prime minister of Great Britain, David Cameron: "Let us be clear"—which you just knew was going to be the rhetorical prelude to his making a partisan point vulnerable to all those objections he was trying to head off.

Storytelling

SUCH RHETORICAL legerdemain was nothing new for Darwin. He had recourse to it more than a few times in the *Origin*. We find it in evidence, for example, where he seeks to persuade us that the eye was not designed but somehow fell into place as the result of a myriad of chance selections over time:

> That many and serious objections may be advanced against the theory of descent with modification, I do not deny. I have endeavoured to give them their full force. Nothing at first can appear more difficult to believe than that the more complex organs and instincts should have been perfected, not by means superior to, though analogous with, human reason, but by the accumulation of innumerable slight variations, each good for the individual possessor. Nevertheless, this difficulty, though appearing to our imagination insuperably great, cannot be considered real if we admit the following propositions, namely,—that gradations in the perfection of any organ or instinct which we may consider, either do now exist or could have existed, each good of its kind,—that all organs are, in ever so slight degree, variable,—and, lastly, that there is a struggle for existence leading to the preservation of each profitable deviation of structure or instinct. The truth of these propositions cannot, I think, be disputed.[61]

What has Darwin said there? According to my reading he suggests that, even though you or I might find unbelievable the idea of almost unimaginably complex structures like the eye coming about by slight and undirected variations over time, the difficulty lies all in our imagination. He then points to three quite doubtful propositions as if they were self-evidently true and as a (hoped for) confirmation of his point, all in the hope that we will come round to his way of thinking. But asserting that a firmly felt instinctive reaction is mere imagination is only that, an assertion, not a demonstration; and labeling disputable points indisputable no more makes them so than praising the proverbial "emperor's new clothes" cures his nakedness.

Once the vulnerability of Darwin's arguments is shown in one instance, the rest of his "story" seems more questionable—rather like when, in criminal cases, if suspects are caught out in one fabrication, their credibility and their testimony collapses like a house of cards. To give one or two more examples: Darwin was never able to give a straight answer to those persons who objected to his explanation of why giraffes had long necks. If this were such a selective advantage, why did other animals not evolve long necks? In fact, why were not all species evolving in all different directions, ostriches acquiring the useful faculty of flying, other terrestrial animals of swimming, and so on?

Such objections were thoroughly, and indeed devastatingly, analyzed by Himmelfarb, but she was not the only critic who had found implausible Darwin's apparent make-it-up-as you-go-along speculations. In 1894 Bateson stated that Darwin had shown the possibility of his theory but not its probability.[62] The fact that so much of the theory is not evidence-based has bequeathed to future adherents of his ideas a very difficult legacy. In fact, some of those supporters have found themselves batting on such a sticky wicket that they have had to resort to the most eye-watering logical contortions to prop up what Himmelfarb had by 1959 concluded was a radically defective theory. Richard Dawkins, for instance, the most indefatigable contemporary expositor of the Darwinian legacy, has at various points in his voluminous publications attempted to cajole us into believing propositions of quite staggering improbability. *Climbing Mount Improbable* was the bluffing title of just one of these volumes, in which he seeks to persuade us that what *we* see as being improbable is not *really* so given natural selection's circuitous route to its summit over cosmic swathes of time. In his combatively titled *The Blind Watchmaker,*[63] he is obliged to fall back on a repertoire of persuasion techniques like those previously employed by Darwin himself (such as logical sleights of hand, *pro domo* modes of presentation of evidence, and special pleading) in the attempt to coax us into believing many things against which logic and common sense scream out in protest.

Like Darwin before him, then, Dawkins has to try to persuade us to abjure our innate instincts of common sense (i.e., the kind of "smell test" scenarios dealt with by Malcolm Gladwell in his bestseller *Blink,* which preserve us from so many mistakes in the course of our daily lives). He challenges us to rise above our "decade-bound imaginations" in order to understand the true dimensions, and therefore creative and transmutational possibilities (in his view), of geological time. Evolution, Dawkins explains in an urbanely reasonable-seeming manner, has equipped our brains to assess probability only in terms of three score years and ten. We are therefore encouraged to believe that our "difficulty" in comprehending the whole microbes-to-monkeys-to-man postulate lies with the huge timescale involved in these metamorphoses: "Evolution has equipped our brains with a subjective consciousness of risk and improbability suitable for creatures with a lifetime of less than one century. Our ancestors have always needed to make decisions involving risks and probabilities, and natural selection has therefore equipped our brains to assess probabilities against a background of the short lifetime that we can... expect."[64]

As Neil Broom observed, at such moments "megatime becomes the instrument of creative change. It is used as a kind of magic wand, waved at appropriate points in the argument in order to accomplish quite remarkable feats of materialistic magic."[65] Skeptics of such a tactic are charged with using the so-called Argument from Incredulity—a mocking term coined to imply that dissenters should gainsay the cardinal virtue of critical rationality and make the ascent to a supposedly higher plateau of insight. However, as Nagel notes, "I believe that the defenders of intelligent design deserve our gratitude for challenging a scientific world view that owes some of the passion displayed by its adherents to the fact that it is thought to liberate us from religion."[66] It is no mere coincidence that Dawkins, among his numerous other books, is also the author of *The God Delusion*.

4. Cosmos and Chaos

Intelligent design, as one sees it from a scientific point of view, seems to be quite real. This is a very special universe: it's remarkable that it came out just this way. If the laws of physics weren't just the way they are, we couldn't be here at all. The sun couldn't be there, the laws of gravity and nuclear laws and magnetic theory, quantum mechanics, and so on have to be just the way they are for us to be here. Some scientists argue that "well, there's an enormous number of universes and each one is a little different. This one just happened to turn out right." Well, that's a postulate, and it's a pretty fantastic postulate—it assumes there really are an enormous number of universes and that the laws could be different for each of them. The other possibility is that ours was planned, and that's why it has come out so specially.

—Physicist and Nobel Laureate Charles H. Townes[1]

Almost a century ago the philosopher Bertrand Russell had some grave words to deliver on the plight of finding oneself in a universe bereft of meaning and metaphysical consolation:

That man is the product of causes which had no prevision of the end they were achieving; that his origin, his growth, his hopes and fears, his loves and beliefs are but the outcome of accidental collocations of atoms; that no fire, no heroism, no intensity of thought and feeling, can preserve an individual life beyond the grave; that all the labours of the ages, all the devotion, all the inspiration, all the noonday brightness of human genius are destined to extinction... that the whole temple of man's achievement must inevitably be buried—all these things, if not quite beyond dispute, are yet so nearly certain, that no philosophy which rejects them can hope to stand. Only within the scaffolding of these truths, only on the firm foundation of unyielding despair, can the soul's habitation henceforth be securely built.[2]

As a literary historian, Russell's semi-macho, crypto-military tone of unyielding defiance triggered for me something of a shock of recognition. The sentiment undergirding those sonorous words would not seem out of place within the warrior ethos informing the heroic poetry of the ancient world or that of the early Middle Ages. Such heroism under adversity is reminiscent of the unflinching bravery commended in the Old English poem which records a somber Anglo-Saxon defeat at the hands of Viking raiders, *The Battle of Maldon* (991 AD): "Mind must be firmer / heart the more fierce, Courage the greater, / as our strength diminishes."

In Russell's own day, an echo of that old heroic ethos had found expression in the influential Nietzschean philosophy of the *Übermensch* (elite human being) who should accept the "death of God" without demur and go forth to triumph over the slave mentality (*Sklavenmoral*) that had held him and his kind in a disenfranchised condition for so many centuries. While the instrumentalization of Nietzsche's philosophy by the Nazis to bolster the case for genocide led to an understandable neglect of Nietzsche after the Second World War, the German philosopher appears to have made something of a comeback with the advent of a group of militant writers sometimes termed the New Atheists. Richard Dawkins in particular admonishes us in crypto-martial tones reminiscent of both Nietzsche and Russell to have the courage to face up to the inevitable void after our deaths and forge our own fearless paths through life in a similar spirit of bleak existentialist heroism, unblinking before the sobering revelation that "the universe we observe has precisely the properties we should expect if there is, at bottom, no design, no purpose, no evil and no good, nothing but blind, pitiless indifference."[3]

Nihilism in the Dock

HOWEVER COURAGEOUS one may find such sermons, whether from Dawkins or Russell or Nietzsche, they could be said to have acquired a kind of sepia-tinted datedness about them.[4] In the last half century, advances in the world of cosmology have revealed that our planet turns out to be biofriendly to a well-nigh miraculous degree—a verdant oasis fine tuned in a dizzying number of ways for life, in contradistinction to the little less than Hadean depths found in possibly the entire remainder of the observable universe. Through the lens of a celestial telescope, it is true, one can see little but the unfeeling immensity of that unremittingly hostile universe invoked by Russell, but if we look around us here on Earth we can see a planet which seems entirely discontinuous with the rest of the observable cosmos and abounding in a host of benign phenomena so numerous that they tend to go largely unnoticed.

Russell's assumption of material forces churning away mindlessly over the eons and at length spewing out the unplanned anomaly of human life—"a curious accident in a [cosmic] backwater"[5] he once termed it—was first formally challenged by astrophysicist Brandon Carter in 1973.[6] Carter put forward what he termed "the anthropic principle" (from the Greek *anthropos*, man). According to Carter's detailed calculations, the fact that our planet is habitable, and exists in a universe that could generate and host a habitable planet such as Earth, obtains thanks only to numerous finely tuned conditions, many of them stretching back to the first nanosecond of the Big Bang. Many of the ways that Earth appears fine tuned for life had been noted previously,[7] but Carter made an advance in formalizing planetary and cosmological fine tuning, and he jump-started a wider conversation in the community of physicists, astronomers, and cosmologists about possible explanations for this fine tuning.

Already in the 1960s scientists had begun to notice a strange connection among a number of otherwise unexplained coincidences in physics. It emerged that many of these mysterious values could be explained by one overarching fact: the values had been necessary for the origin and preservation of human and other life. Some of the fundamental constants referred to include the particular strengths of the electromagnetic force and the force of gravity, which appear to be calibrated with extraordinary precision (to a dizzying number of decimal points) for human needs. The Earth, too, caters to human needs in a host of ways unknown to scientists of a century or more ago. Its magnetic shield, for instance, prevents our atmosphere from being stripped of components crucial for life. This and various other planetary conditions have grown so numerous that some astrobiologists have despaired of finding other habitable planets among the hundreds of millions of stars in the Milky Way,[8] while many others have at least ceased to talk about expecting to find habitable planets around every third or fourth star in the galaxy.

Privileged

WHAT ARE we to make of this radical discontinuity between the Earth—the only location known to have both a geosphere and a biosphere—and the rest of the cosmos? The contrast between our life-promoting biosphere and the unremitting deadness of so much of the rest of the cosmos is a stark one.

It was then for good reason that Carter concluded that our place in the cosmos, while not at the center of the universe as had been supposed prior to Copernicus, was at the very least "privileged." Of course, this sounds a trifle Panglossian, after the fictional figure of Dr. Pangloss, whom Voltaire invented to guy Enlightenment credulities with the character's fatuous refrain that eighteenth-century men and women were living "in the best of all possible worlds." And yet there is no doubting Carter's evidence that we do indeed live at a particularly privileged address in the universe. When one considers the manifold ways that it is fine tuned for habitability, Carter's choice of the adjective "privileged"[9] almost seems an understatement. Some might prefer the term "uniquely blessed"—meant either literally or metaphorically.

Carter's findings, initially announced at a specialist conference, have since been incorporated into the mainstream of cosmological understanding, despite dissenting opinions from some scientists embarrassed by the possible theistic implications of the new discoveries. In a series of books appearing over the past few decades aimed both at subject specialists and lay persons, astrophysicist Paul Davies elaborates on the growing awareness by astronomers that the fitness of our earthly environment for life seems all too great to be accidental, and that the laws of physics appear to be uncannily fine tuned to support humankind.[10] Such evidence runs counter to the older opinion that our place in the cosmos arose by a process of cosmic vicissitude that "did not have us in mind," as American zoologist George Gaylord Simpson opined many decades ago.[11]

Arguably, these factors even go some way toward relativizing that purported demotion of humankind brought about by the so-called Copernican Revolution,[12] a point Carter himself made in his seminal 1973 paper introducing the concept of the anthropic principle. Our solar system is of course indubitably heliocentric rather than geocentric, but recent research excavating the many ways Earth is unusually—and perhaps uniquely—fine tuned for life, combined with discoveries that the universal laws of physics and chemistry also appear fine tuned to allow for the existence of life, and indeed, fine tuned in some ways to allow for advanced terrestrial beings such as ourselves, cannot be simply shrugged off—not if we are serious about following the evidence. For Michael Denton this evidence signals a return to the kind of anthropocentric conception of the world of the pre-Copernican Middle Ages. People in the Middle Ages got many things wrong, he concedes, but their most presumptuous conviction, that of humankind's prominence in the great cosmic drama, seems to have stood the test of time.[13]

One inference from the above is that life may not after all simply be the aleatory consequence of where the cosmic dice had happened to fall. It is at least warrantable in logical terms to infer that a power greater than mere happenstance may have been responsible for the benign dispensation. For that reason Denton begs to differ from modern liberal theologians who hastily resigned themselves to seeing science and theology occupying distinct epistemological realms, "non-overlapping magisteria" [= domains] in Stephen J. Gould's somewhat cumbersome wording.[14] (Many have glossed that expression as a polite euphemism in which the right to identify truth is ceded to science, whereas religion is confined to the more peripheral domain of subjective value.) On the contrary, counters Denton, there is important overlap, for science has provided evidence that the laws of nature (and whatever actuates them) appear specifically devised to support life. The magisteria overlap and urge scientists and theologians to come together in dialogue. On this reading of things, contemporary theologians—and religious people in general—have been too supine before the mighty behemoth of Science and, in the process, have missed scientific evidence they might have alighted upon.

Martian Mirage

The findings of Copernicus and Galileo "de-centered" planet Earth and were later enlisted in what is conventionally referred to as the "principle of mediocrity," meaning that Earth, no longer special, takes its place as just one planet amongst numberless others. Succeeding generations took Earth's "mediocre" status very much to heart. This led to the belief that the worlds of our solar system must be so similar to Earth that they must surely be inhabited by intelligent life. Curiously, this notion at one time was accepted even by many influential Christian thinkers, who could not see the purpose of these other planets unless God had chosen to populate them.

For years humanity seemed to want to convince itself that Mars was an outstanding candidate for advanced life beyond Earth. When in 1877 astronomer Giovanni Schiaparelli observed what he perceived through his telescope as conduits on the Martian surface, he called them *canali*, a word which in Italian means a channel of some sort but not necessarily the man-made construction that the term "canal" indicates in the English language. The ensuing linguistic confusion was enough, at least in the Anglosphere, to unleash eager imaginations, encouraging the idea that a race of extraterrestrial inhabitants was active on Mars—a notion given further currency in the successful science fiction of H. G. Wells. The American astronomer Percival Lowell even wrote a book entitled *Mars and Its Canals* in 1905, but in the first decade of the twentieth century the idea that the *canali* had been engineered by extraterrestrials was opposed by none other than Darwin's old comrade-in-arms from a previous era, the octogenarian Alfred Russel Wallace.

Wallace's *Is Mars Habitable?* was published in 1907, when its author was 83 years old. Wallace argued, using technical knowledge about planet cooling ratios relative to distances from the sun, that Mars was far too cold to allow water to flow into the so-called *canali* and that life was highly unlikely on the planet for the additional reason that the atmosphere was too thin. Surprisingly, despite the work of Wallace and other notable skeptics in the years that followed, the "canal" myth persisted in some quarters, not being laid fully to rest until 1965 when the Mariner 4 space probe sent back close-up pictures of the Martian surface, revealing that the appearance of "canals" in the earlier photo was due to chance alignments of physical features on the planet's surface along with the blurring of the original telescope image by our atmosphere. With the new higher resolution images, the whole "Martian canal" affair quickly took on the retrospective appearance of a collective form of wishful thinking.

Yet hope, or credulity, springs eternal. Even in the present day, searches are ongoing for at least microbial life on Mars—with disappointing results. And in what appears to smack of what the French term "professional deformation," scientists have fallen prey to what Jacques Barzun called the genetic fallacy, which he defined thus: "Because living things depend on certain chemico-physical things, therefore human beings [and by extension, any possible extra-terrestrial beings] are physico-chemical combinations and nothing more."[15] This mindset, along perhaps with an understandable desire to justify additional government funding, go a long way to explaining how some very bright NASA mission scientists can seem to base their searches for extra-terrestrial life on the rather simplistic nostrum of "water + organics = life." In other words, find these ingredients together on a terrestrial-type planet and expect to find life.

For the many of us still awaiting a convincing explanation, this contention might appear to represent an impermissible leap of faith, but it is all part and parcel of that deterministic school of biology which is the prevailing view of American Space science, and one which has spread to inform the views of much of the media commentariat. I was reminded of this way of thinking on watching a recent TV series on cosmology where an expert presenter (Michelle Thaller, assistant director of NASA's Goddard Space Center) expressed a somewhat wistful sense of kinship with rocks because, as she put it, rocks and humans are both carbon-based!

A form of what might be termed materialist credulity appears to afflict this guild of mathematical geniuses who interpret the skies for us. The same presenter also confessed that recent cosmological advances had not identified the habitable planets (let alone their postulated denizens) that she had once confidently hoped would be found in great abundance in our corner of galaxy, and beyond. She concluded rather plangently that the community of astronomers "were expecting heaven, and instead found hell." A co-presenter of the same program, Professor Michio Kaku of the State University of New York, confronted by the same apparently limitless swathes of uninhabited and uninhabitable space, put the same thought more pithily: "Boy, were we wrong!"

The belief in the existence of extra-terrestrial civilizations had already been challenged (but clearly not vanquished) by one of Darwin's Cambridge tutors, William Whewell, who in his *Of the Plurality of Worlds* (1853) attempted to demonstrate that the Earth is special and that life is unlikely elsewhere in the cosmos. Anticipating the anthropic principle by considerably more than a century, he wrote, "The Earth... is the abode of life... because the Earth is fitted to be so, by a curious and complex combination of properties and relations, which do not at all apply to the others."[16] Whewell considered that, in any case, the existence of intelligent life on other worlds was incompatible with humankind's special relationship to God.

Whewell was able to buttress the non-theistic portion of his argument by appealing to the latest geological discoveries of Lyell, showing that for long tracts of time our Earth had remained uninhabited. This proved that worlds uninhabited by sentient persons are not impossible, having a precedent in the ancient geological condition of the once-barren Earth itself. Whewell's prescient contentions have been taken up by other scientists such as David Waltham, who in his book *Lucky Planet* acknowledges the "striking similarity" of many of Whewell's arguments to his own—except that Waltham balks at accepting the reason Whewell gives for our good fortune: divine providence.

In place of providence, Waltham, referencing the title of his own book, simply puts our good fortune down to luck: "a good fortune that is inevitable somewhere in a big enough universe."[17]

Cosmic Lottery

IT MAY be, however, that the odds are too long even for our vast universe with its billions of galaxies, so many and so unlikely are the many conditions Earth enjoys that are essential for long-term habitability. And in any case, all the stars in the universe, and all their myriad planets, can do nothing to explain the fine tuning of the laws and constants of physics and chemistry for life. We have our one universe, and its laws and constants are finely tuned to allow for the formation of stars, the synthesis of the life-essential elements in the core of stars, terrestrial planets, and on and on. Why are the laws themselves fine tuned?

In the last few decades some scientists have sought to solve this riddle by advancing the theory of a whole ensemble of universes, the idea being that there may be countless other universes parallel to our own, each with different natural laws, and if so, the somewhat idiosyncratic logic goes, then the series of miraculous coincidences that produced life in our world are to be expected since at least one of those universes had to be the winner of the great cosmic lottery. And of course, having granted such a multiverse, we shouldn't be surprised to find ourselves inhabiting a universe and planet commensurate with intelligent life. After all, if we weren't, we wouldn't be here to notice our good fortune.[18]

There are all sorts of things to observe about this curious hypothesis. Perhaps most fundamentally, as Rupert Sheldrake and others have pointed out, the idea does not in fact logically exclude God: it simply increases the divine domains.[19] Indeed, as Stephen Meyer and others note, even if one grants the multiverse hypothesis and a multiverse-generating mechanism, there are strong reasons to conclude that the multiverse-generating mechanism itself would have to be fine tuned in order to produce even a single universe compatible with life.[20]

More than a few prominent physicists are deeply uncomfortable with the multiverse hypothesis, positing as it does a myriad of unobservable universes to solve a persistent appearance of deliberate fine tuning, a hypothesis borne along by a rose-tinted credulity that benign outcomes can be expected by accident given enough time/opportunity/space. Being unobservable and largely untestable, this postulated ensemble of other universes cannot aspire to any empirical status even at any foreseeable time in the future.

This appeal to luck and an imagined multiverse seems to me at least as much of a cop-out as Dawkins's belief in time as the eventual bestower of all benign outcomes in the realm of biological evolution. For me at any rate, Waltham's conjecture is less convincing than Whewell's old providentialist argument if measured against strict standards of logical probability, counter-intuitive as that will doubtless seem to many—including, at some level, myself.

Proxy Wars

ONE OF Darwin's most feared opponents, St. George Mivart, once noted, "If the *odium theologicum* has inspired some of its (Darwinian) opponents, it is undeniable that the *odium antitheologicum* has possessed not a few of its supporters."[21] By this he meant that both Darwin's supporters and his detractors were biased by their religious or else anti-religious preconceptions and that positions taken up pro or contra Darwin were in an important sense proxies for profounder ideological beliefs. Mivart was both a distinguished scientist and a (Catholic) theist, and we may surmise that, although most of his objections to Darwinism undoubtedly rested on scientific foundations, it is not possible to discount the prima facie possibility that some of his opposition was religiously motivated. What is equally clear, however, is that the opposing camp was motivated in part at least by the desire to champion the opposing cause of philosophical materialism. A contemporary German scientist, August Weismann, admitted this with some candor in an essay published in 1893:

> We must assume natural selection to be the principle of the explanation of the metamorphoses, because all other apparent principles of explanation fail us, and it is inconceivable that that there could be yet another capable of explaining the adaptation of organisms, without assuming the help of a principle of design.... We accept natural selection not because we are able to demonstrate the process in detail, not even because we can with more or less ease imagine it, but simply because we must, because it is the only possible explanation that we can conceive.[22]

Nor is Weismann simply an historical curiosity. More recently, Harvard geneticist Richard Lewontin echoed Weismann while framing his dogmatic commitments even more broadly to encompass not just Darwinian materialism, but materialism generally: "We take the side of science *in spite* of the patent absurdity of some of its constructs... *in spite* of the toleration of the scientific community for unsubstantiated 'just-so' stories, because we have a prior commitment, a commitment to materialism."[23]

If we are to call a spade a spade, the above statements can only be read as exhibiting an extraordinary degree of intellectual dishonesty, announced with uncontrite *chutzpah*.

There is here, too, a Great Men of Science narrative, one committed to casting Darwin alongside Copernicus, Galileo, and Newton, and offering him as a beacon of the scientific Enlightenment, a poster boy for the Olympian sentiment of scientific progress, *citius, altius, fortius* (faster, higher, stronger). The paranoid fear seems to be that if Darwin were to be toppled—or even seriously questioned—this would involve the rejection of the scientific method *tout court* and a return to a dreaded theocratic ordering of things.

Whereas pre-Darwinian generations believed unselfconsciously that nature's laws were God-given, post-Darwinian scientists have the greatest difficulty with such discourse. In the post-Darwinian era naturalistic explanation alone was valued and metaphysical speculation shunned, since "science is generally thought of as excluding from its scope any higher meanings. When it ceases to do so it ceases to be science and becomes philosophy."[24] Hence the majority of science professionals reject any kind of "God-talk" in the academy, the more so if it is suspected of being camouflaged for tactical reasons; hence the frequent imputation of "neo-creationism" to proponents of intelligent design.

How justified is the imputation? Easy generalization is clearly not possible where great variety, and disparity, is involved; and what is referred to for convenience as the intelligent design movement comprises a disparate array of scientists of different nationalities, religious backgrounds, and worldviews, united only by their conviction that certain features of the natural world are best explained by reference to an intelligent cause or causes rather than to any purely blind process, such as natural selection.

There are, of course, some Christian groups, especially in the United States, who seek to use the scientific controversy for their purposes. They present a mirror image of attempts to instrumentalize Darwinism for the cause of militant atheism. However, among those pointing to evidence of design in the natural world, religious motivations are not universal; and in any case, such motivations can and should be distinguished from the scientific evidence and arguments made by design proponents. A willful failure to do so is to embrace the *ad hominem* fallacy.

Even if one is unwilling to cease motive mongering, there are those against whom it is very difficult to wield such a rhetorical weapon. Paul Davies, for instance, explains in his book *The Mind of God: Science and the Search for Ultimate Meaning* (1992) that he is not religious but that "through my scientific work I have come to believe more and more strongly that the physical universe is put together with an ingenuity so astonishing that I cannot accept it merely as a brute fact."[25] In other words, the scientific evidence, rather than any religious motivation, led him away from regarding our finely tuned universe as only a mindless accident. The late humanist philosopher and doyen of the British Rationalist Association, Professor Antony Flew, described a similarly science-based motivation in his book *There Is a God: How the World's Most Notorious Atheist Changed His Mind* (2007). For decades, he had been a world authority on the philosophy of religion and an influential atheist philosopher, but as he explained in the book, his conversion from atheism was precipitated by his grappling with fresh scientific evidence not available in his youth.

This evidence, Flew explained, stemmed largely from the sheer intricacy of microbiological structures, which can only now be fully appreciated by analogy with the miniaturizations enabled by advanced nanotechnology. He wrote in 2010, "It now seems to me that the findings of more than fifty years of DNA research have provided materials for a new and enormously powerful argument to design."[26]

The invention of the electron microscope in 1946 enabled insights into the microworld unimaginable to the world in which he grew up. Molecular biology has increasingly shown that the microscopic and sub-microscopic worlds are as extensive in their own terms as that vaster world of extra-galactic structures and exoplanets. It sometimes appears to the present author as an ever-smaller, seemingly endless regression sequence of Russian dolls, worlds within worlds, the number visible depending only on the magnifying powers of the devices used to study them.

Individuals have long been aware that the works of nature are, in William Paley's words, greater than those of mankind "in a degree which exceeds all computation."[27] Douglas Dewar, the largely forgotten British scientist who opposed Darwinian doctrine between the two World Wars, wrote: "The simplest cell, the unit of every organism, has a structure compared to which that of a modern printing press or a watch is simple and clumsy."[28] However, advances in the last half century or so have been so great that neither Paley in the 1800s nor even Dewar in the 1930s could have imagined the more recent discovery of what was tantamount to a whole undiscovered continent in the molecular realm. Being told that nature's protein-synthesizing apparatus is of the order of several thousand million million times smaller than the smallest piece of functional machinery ever constructed by man from scratch would, I believe, lead most people to scratch their heads over who or what could possibly have been behind something so small and intricate.

This newly discovered micro-universe was what Flew was responding to. He was more than willing to concede that his present position could have religious implications, but quite reasonably pointed out that it did not depend on religious presuppositions. Those at the time who resorted to cod psychology to suggest that Flew's advancing years made this a kind of pre-deathbed conversion were in my opinion not simply being crass and in lamentable taste but—far more to the point—very wide of the mark. Flew described his striking *metanoia* as "a pilgrimage of reason," and, notwithstanding the apparent historical incongruity of persons finding their way to God by reason rather than by faith, his move might justly be characterized as an instance of an as yet little-studied phenomenon: scientific conversion.

In any case, to set the record straight, Flew is not known to have become a Christian; he merely accepted the existence of God. He described his new position as a form of deism, accepting that there is supernatural intelligence responsible for creating the universe, but he rejected special revelation in the form of the Bible and the notion of a personal God in the sense of one who watches over his terrestrial flocks. He expressed an openness to the possibility of special revelation, but made it clear that his present view was that God does not intervene, and has not intervened, in human affairs since Creation. To quote Flew in his own terms:

> I must stress that my discovery of the Divine has proceeded on a purely natural level, without any reference to supernatural phenomena. It has been an exercise in what is traditionally called natural theology. It has had no connection with any of the revealed religions. Nor do I claim to have had any experience of God or any experience that may be called supernatural or miraculous. In short, my discovery of the Divine has been a pilgrimage of reason and not of faith.[29]

Flew and other such intellectual dissenters, many of whom come from the same academic ranks as the Darwinian orthodox, should not in my opinion be lightly doubted or have their *bona fides* discounted because of prejudice. It would be anachronistic to frame this modern debate as "Enlightened science *versus* unexamined Creationism, Part 2" (the binary division best known from the infamous Scopes "monkey trial" of 1925 in Tennessee—and, more precisely, from the famous and almost cartoonishly rendered Hollywood film depicting the trial, *Inherit the Wind*, which flattened out important nuances and grossly distorted important elements of the story).[30] Such is unfortunately the kind of knee-jerk reaction of those who oppose any dissent from Darwinian orthodoxy. This kind of misrepresentation of an opponent's position does nothing to help the cause of rational public debate (and is, incidentally, demeaning to those who advance such misrepresentations). Imputations of bad faith (e.g., charging Darwin skeptics with illicit religious motivations) or strawman characterizations of their arguments (e.g., labeling carefully considered doubts about Darwinism as "arguments from incredulity") are valueless, except in deflecting attention from the gravamen of the design theorists' evidence and arguments.

Denton, Flew, Behe, Dembski, and other prominent defenders of intelligent design focus squarely on the substantive issue of the theoretical and empirical inadequacies of the Darwinian paradigm, and on what they see as the superior explanatory power of the design inference. Referencing Thomas Kuhn's work on paradigm shifts, they see Darwinian theory as persisting from professional and cultural inertia rather than on the basis of its present explanatory merits.

To be sure, intelligent design may not be science in the sense that it can develop new knowledge by hypothesis-testing, followed by modification of the original theory based on experimental results and then renewed testing as a "control" to the preceding experiments—but on the other hand Darwinian natural selection has no testable scenarios either, and in the very nature of things never can have—unless a particularly convincing cache of those ever-elusive missing links should be disinterred. In my own view, design represents a more compelling inference to the best explanation.

5. THE MYSTERY
OF MYSTERIES

Scientists should never present themselves as certain when there is doubt. The very success and truthfulness of science is founded on doubt and scepticism. It moves forward by continually rethinking, reobserving and rechecking against reality again and again to expose the flaws in current ideas.
—Philosopher of evolutionary science Fern Elsdon-Baker[1]

Noam Chomsky once remarked that all matters of human puzzlement can be subdivided into two categories: problems, which can be solved, and mysteries, which cannot. He came to the conclusion that our minds must inevitably experience "cognitive closure" with regard to some of the profoundest topics of human enquiry. He cited free will as one such mystery and human consciousness as another.[2]

Darwin himself acknowledged our inability to unravel every mystery when he stated that some things were unfathomable to us and that we had no more chance of understanding them than a dog would the mind of Newton.[3]

For this reason there is no shortage of awkward questions which bright young people ask and adults can't answer. These are the existential "biggies" which people of all ages would love to have answered but in the face of which experimental science tends to fall silent, such as: How did a once barren terrestrial environment give rise to life forms? How did the resources deemed necessary to this process—self-replicating molecules bearing genetic information—arise in the first place? What is the ultimate origin of the genetic code, and who or what directed it to produce plant and animal species? Why are we safely cocooned in a cosmic Goldilocks zone when so many parts of the universe are more reminiscent of

Dante's *Inferno?* Where do the laws of physics come from? What was before the Big Bang? Why is there something rather than nothing?

The laws of nature do not, alas, answer any of the above. Physics can explain much of the physical universe but not the laws of physics themselves. In fact, scientific laws do not explain the world to us even to the degree one might imagine; they merely describe certain regularities in mathematical terms, and are often referred to for that reason as secondary causes, in contradistinction to the first cause—historically referred to as the *causa causarum*, the ultimate cause of all things.

Sir Isaac Newton made this point after his discovery of the law of gravity, which, he remarked, "explains the motions of the planets, but cannot explain who set the planets in motion" (or for that matter, he could fairly have added, who or what established gravity itself). Newton's law of gravity, in other words, did not create gravity.

Oxford mathematics professor John Lennox, in his book *God and Stephen Hawking*, reminds us that the laws of physics are descriptive and predictive but not creative in their own right. That is, as mathematical descriptions of natural regularities, they are wholly unable to produce anything. Lennox illustrates with a modern-day parable:

> In the world in which most of us live, the simple law of arithmetic itself, 1+1=2, never brought anything into being. It certainly has never put any money into my bank account. If I put £1000 into the bank, and later another £1000, the laws of arithmetic will rationally explain how it is that I now have £2000 in the bank. But if I never put any money into the bank myself, and simply leave it to the laws of arithmetic to bring money into my bank account, I shall remain permanently bankrupt.[4]

Thus, the laws can explain how the jet engine works, but not how either the laws or the first jet engine came to exist. The reasonable answer to how the jet engine originated would not be a description of the underlying physics but rather, "Frank Whittle and a team of highly skilled aeronautical engineers invented it."

Scientists may try to get rid of the notion of a Creator but are then obliged to confer quasi-creatorial powers on precisely these mathematical descriptions that go under the name of laws. To do so is to fall deep down Alice's rabbit hole. Indeed, when Galileo famously stated that the "the laws of nature are written by the hand of God in the language of mathematics," he might have been more than a little bemused to have been told that a future scientific *confrère* would even moot the possibility of mathematical laws writing themselves.

Ultimate questions must always be beyond the scope of empirical science as conventionally defined. Given this, it might have been truer had Darwin applied the expression he used about the origin of life to its sequel as well and simply stated that the whole question of biological origins was a work in progress and, as to definitive conclusions about ultimate causes, these must remain, to repeat Darwin's Latin tag, "*ultra vires* [beyond our powers] in the present state of our knowledge."[5] Such a concession would certainly have been more consistent with his riven state of mind on this issue, a conflict which endured long after the original publication of *Origin*, evidenced by his wrestlings in his *Autobiography*, penned in 1878, where he argues against a benevolent God, then calls himself a theist, and then two pages later calls himself an agnostic. (More on that below.)

The Origin of Species itself is full of such phrases as "the laws impressed on matter by the Creator."[6] Similar expressions can be found in his first pencil sketch of his theory written in 1842 and read only by himself, a context in which there was little if any motivation to throw a sop to a religious public.

There is a clear indication that the Darwin of the *Origin* and later, even in his more Creator-friendly moments, much preferred scientific explanations that restricted God's creative activities to the beginning moment of Creation—the distant cosmic watchmaker of deism rather than the hands-dirty-in-the-garden god of Judeo-Christian theism. The concluding sentence of *The Origin of Species* certainly has a deistic flavor. There he describes his theory of evolution as a "view of life, with its several powers, having been originally breathed into a few forms or into one; and that, whilst this planet has gone cycling on according to the fixed law of gravity, from so simple a beginning endless forms most beautiful and most wonderful have been, and are being, evolved."[7]

Darwin's conflicted thinking, and conflicting pronouncements, on questions at the intersection of science and theology can be exasperating for those wishing to figure out precisely what his theological conceptions were. While some of the inconsistency can be chalked up to public posturing, he appears to have been genuinely ambivalent on the subject throughout his life. According to Neal Gillespie, this "epistemological double vision" resulted from the fact that "early in his career he largely dropped theology from his science but not from his world view."[8]

An Evolving Theory

Darwin's vacillating extended as well to the purely scientific aspects of his theory and argument. After first publication in 1859, the *Origin* was eventually to go through five more, often heavily revised editions as Darwin wrestled with a wealth of scientific objections to his theory. As intransigent as Darwin might have seemed in pressing many points in the *Origin* (or in some cases, cunningly insinuating them, in a way that recent Darwin biographer A. N. Wilson rightly terms "slippery"), he nevertheless in time agreed to relent on a number of important issues. As early as 1862 Darwin entertained doubts about the theory of natural selection as the sole determining factor in the process of evolution, and began to seriously consider the possibility of plural causation. Appealing now to early nineteenth-century evolutionary theories from which he had heretofore tried to distance himself, like those of Buffon and of his grandfather, Erasmus Darwin, he was now prepared to moot the possibility of the environment triggering changes, and he even flirted with the Lamarckian idea of the relative use/disuse of organs as determinants of survival. In the sixth edition of 1872 Darwin expressly stated that "I formerly underrated the frequency and value of these latter forms of variation."

In such ways did he make the honorable attempt to incorporate responses to criticisms leveled at him by other scientists, which meant that the last, heavily emended version of the *Origin* was markedly different from its 1859 original. All these changes, however, took a toll on the clarity of his exposition. As Jacques Barzun put it, the final edition was marked by "self-contradiction, hedging, endless shuffling with words, vacillation."[9]

We should not, however, make too much of this hedging and shuffling. Some of this was undoubtedly of the tactical retreat variety, the better to turn and take a stand on what he hoped was firmer ground after the initial decade of scientific criticisms had been absorbed. Darwin had struck upon a purely naturalistic explanation for the origin of species, and he was not prepared to forfeit the naturalistic element of that theory. God might be allowed a foot in the great entryway door to the cosmos, at the moment of Creation—might be allowed the possibility of existing as some distant prime mover. God might even, in Darwin's more indulgent moments, be allowed to breathe life into the first living organism (or organisms), as he suggests in the closing sentence of the *Origin*, but having articulated his theory of evolution, Darwin's vacillations between agnosticism and deism seem never to have seriously drifted so far as to allow God to muck about in biology proper. As we saw in Chapter 1 above, and as he put it to Charles Lyell, "I would give absolutely nothing for theory of nat. selection, if it require miraculous additions at any one stage of descent."[10]

Soon enough there was a chorus of voices who would give nothing for the theory, period.[11] Eminent American biologist Vernon L. Kellogg called for the Darwinian theory to be rejected,[12] and as noted previously, a substantial counterargument was published by Henri Bergson, in a book honored by the subsequent presentation to its author of the Nobel Prize.[13]

But the pull of the grand solution to the mystery of mysteries proved too strong, and upholders of the Darwinian vision would not be deterred. In the following two decades influential figures in the scientific world scrambled to invigorate the Darwinian legacy by bringing it up to date via a synthesis with Mendelian genetics, consecrating the resulting accommodation as "neo-Darwinism" in the early 1940s.

Since that time, proponents of the overhauled doctrine have fought tooth and nail to defend it against principled objections from scientific colleagues who had begun to view the whole Darwinian edifice as an offense not only to best scientific practice but even to common sense. Why not throw in the towel? If you bring to origins science an anything-but-God mindset, then you will cling tenaciously to the one purely materialistic theory you believe has any chance, however slender, of explaining the mysterious origin of life's diversity. Darwin's proverbial bulldog, Thomas Huxley, is the textbook case. He disbelieved the theory of natural selection and yet praised Darwin to the rooftops because he saw it as far and away the best tool at hand for driving design from science, and religion from society.

Others who are equally committed to a purely materialistic explanation for the origin of life's diversity have taken a more flexible approach, willing to consider various substitutes and augmentations to the theory of evolution by natural selection—following the venerable model set down by Darwin himself as he complicated his theory over a series of five subsequent editions. But none of these materialistic alternatives have won the day, since each faces problems as great or greater than that facing plain vanilla neo-Darwinism.[14]

Second Thoughts

DARWIN'S GREAT stream of followers would eventually bifurcate. Some would seize on his deism-friendly remarks and build a case for theistic evolution, which might be more accurately described as deistic Darwinism. Others would seize on his more materialistic statements and weave his theory of evolution into a larger evolutionary story that dispensed with divine creation altogether, even at the origin of life and, further back still, the origin of the universe.

But for all the cocksureness of tone of this latter Darwinian tributary, one finds at times hints of Darwin's own reticence, of an awareness of features of the natural world that sit uneasily with the materialist framework. Atheist Francis Crick, according to his surviving son, had seen one of his main aims to be a battle against "animism," a coded reference to the divine. And yet Crick could lapse long enough to speak of nature as being "functionally not dissimilar to a miracle."[15]

Sir Fred Hoyle advanced the thesis that the bio-friendly nature of our world appeared to be a "fix" or "put-up job," as if "a super intellect had monkeyed with the laws of physics"[16] for the purpose of spawning and supporting life. (At the same time, however, Hoyle was capable of issuing his famous panspermia "howler" in the attempt to sidestep any invocation of the deity.)

Even the arch-materialist Richard Dawkins could write that life is "almost unimaginably complicated in directions that convey a powerful illusion of deliberate design," and in the sixth chapter of his *Blind Watchmaker* there occurs this rather disarming comment: "Does it sound to you as though it would need a miracle to make randomly jostling atoms join together into a self-replicating molecule? Well, at times it does to me too."[17]

Cumulatively, such instances may prompt one to wonder how certain of their ground they really are.

Just Like That

To SUM up: a century and a half ago, in place of traditional, theistic explanations in harmony with the Genesis narrative, a theory was advanced which depended for its point of departure on the prior operation of a single, preternaturally fortunate chemical reaction so seemingly impossible in material terms as to be thus far unreproducible in the lab, even allowing for human engineering. The putative reaction is held to have created a simple organism that metamorphosed over time into a succession of progressively more complex plant and animal species. Humankind had not come about by dint of any agency, at least not since the initial creation of the universe, but rather arose by a form of occult automatism resulting from incalculably numerous sequences of biochemical lucky chances sifted by natural selection. In contradistinction to all prior evolutionary speculations, the new theory specifically denied the immanent presence of any motive force guiding biological evolution. And if we are to follow that Darwinian tributary that has come to flood and dominate the academy, we must further conclude that the universe quite literally created itself.

If you are in any doubt that this is indeed the received evolutionary narrative, you have only to look at the statement of a close intellectual ally of Richard Dawkins, Peter Atkins, who referred to the creation as being an "agentless act"[18]—a grand contradiction in terms which even the words' Latin roots proclaim. Things do not happen as if by magic or "just like that," as magician *extraordinaire* Tommy Cooper used to demonstrate on his rib-tickling TV shows by his unmasking of run-of-the-mill conjurors' harmless deceptions. The main butt of Cooper's humor was not so much himself for his (deliberate) bloopers in his character as the poor magician *manqué,* but rather his audience's credulity (or at least half-belief) in supposing that things *could* happen by magic. He was essentially getting us to laugh at ourselves for entertaining such a foolish notion, his *shtick* being essentially a lower-case form of the Brechtian alienation effect.

I sometimes think of Cooper's mocking catch-phrase "just like that" as being a snappier (and more honest) synonym for the more portentous "natural selection." For as nineteenth-century Princeton theologian Charles Hodge saw the matter, natural selection was "a blind process of unintelligible, unconscious force, which knows no end and adopts no means."[19] It's no wonder the matter is rarely put this baldly, of course. If you want to skirt the damning admission that your theory is in any sense magical or metaphysical, you will have to resort to euphemistic periphrasis.

Yet however circumspectly and portentously phrased, a theory which replaces agency and teleology with fortuity and unexplained automaticity might still, whatever its packaging, be termed magical thinking by unbiased adjudicators. Others not bound by the etiquette and proprieties of academic debate might simply roll their eyes and issue that *non plus ultra* of demotic put-downs, "Yeah, right."

Advancing tenuous theories to purportedly explain things which are not only beyond all human comprehension but which are not even to the scale of human comprehensibility is a procedure conspicuously at variance with Einstein's more dignified acknowledgment that "the laws of nature manifest the existence of a spirit vastly superior to that of men, and one in the face of which we with our modest powers must feel humble."[20] Apropos of that glaring disproportion, James Le Fanu recently commented on the "marked discrepancy between the beguiling simplicities of evolutionary theory and the profundity of the biological phenomena it seeks to explain."[21] The author, a medical doctor and writer, challenges us in the subtitle of his book to ponder "how science rediscovered the mystery of ourselves," and draws attention to issues to which most of us, myself included, I have to confess, had previously paid little heed. His intervention merits a short consideration of its content, since it has an important bearing on the frustratingly Sisyphean task faced by all would-be theories of evolution.

Such Creatures!

EVERY DAY, writes Le Fanu, we are the unwitting beneficiaries of an incalculable number of invisible natural wonders, such as the purifying function of the liver (which is able to perform more functions than the largest chemical refinery), or the autonomous functioning of the heart, whose diminutive size belies its enormous pumping power (artificial heart machines, being the size of a chest of drawers, have to be hauled around on trolley wheels and can only be used for a number of hours as a stop-gap before transplant), or our physiological capacity for bodily self-repair: think (I might add) of a bicycle puncture mending itself automatically or the implication of the April Fools ad put out by BMW some decades ago to the effect that the special paints applied to the company's cars were self-cleaning. What seems amusingly preposterous in the case of human manufacture is perfectly practicable in human physiology via the cleansing function of the liver.

We might also usefully ponder the fact that there exists a diminutive universe in each of our individual cells which went all unsuspected for millennia before mid-twentieth century advances in electron microscope technology. Or what about that bodily system we all refer to glibly (but uncomprehendingly) as the immune system—how many people know how that works? No, I didn't either. This is what Linda Gamlin says about its well-nigh preternatural complexity, and keep in mind that the excerpt below only scratches the surface of the immune system's complexity:

> Faced with the onslaught of microbes, how does the normal human body defend itself and stay healthy? To begin with, it keeps out as many potential pathogens as possible with barriers such as the skin and other non-specific defences. The skin, which is waterproof, is impenetrable to most invaders, and it provides fatty acids that many microorganisms find toxic. Areas not covered by the skin, such as the eyes, mouth, lungs and digestive tract, are more vulnerable, but they have alternative defences. Tears, saliva, urine and other body secretions contain lysozyme, an enzyme that can kill certain types of bacteria by splitting the molecules found in their cell walls. Mucus in the nose and airways engulfs bacteria and stops them penetrating the membranes. Cilia—tiny beating "hairs"—then push the mucus out of the airways into the throat, where it is swallowed. In the stomach, acid kills most of the microorganisms in food, as well as starting the process of digestion.[22]

Needless to say, our bodily organs are quite beyond the abilities of human bio-engineers to reproduce. To give a prime example, experiments in the United States to introduce artificial hearts to patients had to be withdrawn some decades ago when the fatalities topped 200 with no realistic hope of medical experts being able to improve their technology. Forty years of research and development and forty billion dollars went down the drain. If such gargantuan efforts and expense could not fashion a functioning heart-substitute, it becomes all the more difficult to imagine a heart being constructed by the serendipity of random mutations and natural selection.

Curiously, many complex wonders of nature have been, as it were, hiding in plain sight for decades, even centuries. Nature was once perceived to be alive with signs and portents by our medieval and even Renaissance forbears. Shakespeare's use of the technique of pathetic fallacy[23] credits nature with a form of indwelling spirit which could actively communicate its meanings to humans.[24] By contrast, the post-Enlightenment centuries brought with them a steady "disenchantment" of the natural world, and a tendency to regard our natural surroundings as prosaic "givens" of little account.

Such dullness before nature is the more culpable today, writes Le Fanu, because, unlike previous generations, we now know of the deep biological complexities that underpin the effortless simplicities of our daily lives. We should therefore be more appreciative of nature's ingenuity and the sheer ease with which we see, hear, talk, eat, drink, make love, and reproduce our kind. Such should be the central core of school biology lessons, promoting a sense of wonder in the young mind at the very fact of existence. The reason that it does not form that core is that scientists and the educational establishment subscribe to the materialistic-mechanistic model of human functioning, and therefore tend not to "do" wonder.

As a footnote to Le Fanu's argument, I would add that we also betray collectively some element of false entitlement about our innumerable boons: we feel hungry, eat, and by some magical alchemy of which we know little—and couldn't care less about—our bodies transform the food into the very substance of our physical frames. We take it for granted that we are born with hinged bones to provide low-friction articulations, eye-protectors (eyelids), tears secreted by the lachrymal glands to lubricate the eyes so that they don't feel scratchy, and an optic nerve to transmit electrical impulses to the brain to decode visual cues so that we can know where we are. We shrug off as unremarkable the fact that broken bones will, unlike broken vases, mend, or the fact that minor wounds will heal by the process to which medical people refer with a complacent lack of affect as "bodily regeneration."

As far as external nature is concerned, we are the beneficiaries of plants' photosynthesis, the process by which plants convert light energy into chemical energy and produce oxygen, yet we give little thought to this bedrock of our existence. (Nobody, by the way, has the first idea about how photosynthesis might have evolved.) The same goes for the sun's warming rays and all those cosmological constants described above. As for that huge symbiosis by which all life is connected productively in a web of interrelated functions (rainwater for crops, grazing animals fertilizing the soil with dung, worms aerating the soil so that crops can grow, and on and on), this is just another part of what we see as our entitlement, assuming we even bother to think about such things at all. The list could be extended practically without limit.

The Miracle Worker

"Can nothing be simple?" has been the question often posed by biology students. The answer is, frankly, no, and the only honest response to the extraordinary intricacies of "simple" nature must be wonder. Our medieval ancestors had a keen awareness of what at the time were referred to as *mirabilia* and which seventeenth-century science referred to unselfconsciously as natural miracles.

Life on earth has traditionally and for good reason been termed the mystery of mysteries, and there is much to ponder in the contention of nineteenth-century Harvard professor Louis Agassiz that life's mysteries were no nearer to being solved after the publication of *The Origin of Species* than they had been before it. Needless to say, neither Darwin nor Wallace would have been minded to see things in that way, because both were responding to their own internalized challenges in the competition to find something no one else had been able to find in generations of evolutionary thinking: a cogent explanation for the diversification of life that made no appeal to a directly interventionist natural theology.[25] Both men pressed on, acting as much in the spirit of conquistadors as discoverers. (Darwin famously rushed out the publication of *Origin* when he feared Wallace might pip him to the post.)

Their goal, albeit not of course explicitly acknowledged (in deference to that laudable Victorian code of gentlemanly reticence which has fallen into disuse in the last half century), was to become recognized as the Lyell of Biology. They thereby hoped to establish the prestige of a discipline which, they were determined, should slough off old-fashioned and discredited biblical notions and so place biology within the prestigious sphere of "pure" science. By analogy with Lyell's geological work, which had rendered the biblical flood superfluous, the quest of Darwin and Wallace was to render the Christian God superfluous to the rolling out of the universe after the moment of Creation, or at least after the appearance of the first self-reproducing organism. But in order to advance their new paradigm, they were obliged to transfer agency to the process of natural selection which, unfortunately, contained within it an insurmountable problem, as Wallace later acknowledged.

Actually there were two problems. The less fundamental but hardly trivial problem was the lack of empirical evidence for the power of natural selection to generate new forms. Darwin appealed to the success of plant and animal breeders to fill the evidential gap, but even setting aside that artificial selection is purposive and therefore an odd stand-in for a mindless process, there is also what Bishop "Soapy Sam" Wilberforce underscored. Contrary to the common caricature of him as an unctuous and obscurantist buffoon, Wilberforce was an Oxford First in Mathematics with a keen interest in natural history and a good working knowledge of animal breeding methods. In his 1860 review of the *Origin*, he noted that domesticate breeders never succeed in breeding a fundamentally new animal form, and what progress they do make always comes with trade-offs. "The bull-dog gains in strength and loses in swiftness; the grayhound gains in swiftness but loses in strength," wrote Wilberforce. "Even the English race-horse loses much which would enable it in the battle of life to compete with its rougher ancestor."[26]

The more fundamental problem is that the "agency" of natural selection invoked by Darwin and, less expansively, by Wallace,[27] was held to operate unselectively, with no notion of purpose permitted to obtrude into the multiple revolutions of its biological lottery. It's not unlike the bonkers situation of a car salesman marketing a car without any engine underneath its hood, the fellow assuring his customer that the car would nevertheless function perfectly well.

Darwin wanted to have his cake and eat it too. Natural selection is a mindless process; Darwin was adamant about that. Yet he habitually repaired to purposive terminology in his descriptions of it, as when he limned natural selection's construction of the eye: "We must suppose that there is a power always intently watching each slight accidental alteration of the transparent layers [of the eye]; and carefully selecting each alteration which, under varied circumstances, may in any way, or in any degree, tend to produce a distincter image."[28]

One might legitimately ask, how it is possible to "intently watch" and "carefully select" unintelligently? That is entirely discrepant with what Darwin elsewhere claimed for the process he invoked. The contradiction points to a more than trivial conceptual confusion, and I would surmise that the very phraseology Darwin uses reveals that he must have had some awareness of the illogicality of his own position, even if only at some barely conscious level of apprehension.

In the course of my professional life in higher education, an area that proudly regards itself as inclusive, it has often struck me as disturbingly contrary to that ideal that some colleagues in science departments betray an almost apartheid-like antipathy to philosophers, with philosophy coming second only to theology in the demonological hierarchy of some of their number. My vague understanding of this antipathy has, however, become much clarified in the course of preparing this volume. For it is philosophers in particular who have typically been the ones responsible for calling out many "mad genius" ideas put forward by representatives of the scientific community—a task for which said philosophers have received few thanks, needless to say—and which doubtless explains something of the animus towards them.

The problem with "natural selection" for philosopher Antony Flew was that it no more resembles any kind of conscious selection procedure than "Bombay duck is a species of duck."[29] It has been described as a would-be materialistic although in reality miraculous explanation. As Le Fanu put it, "Darwin's explanation was in its own way profoundly "metaphysical" in that it attributed to natural selection powers that might reasonably be thought to be miraculous—that it should somehow fashion perfection from a blind, random process, and transform one class of animal into another."[30]

Again, we come up against the difficulty of scientists having to impute creative powers to phenomena with no creative capacity, rather like the way Richard Dawkins anthropomorphizes genes as being "selfish"—apportioning perception and decision to inanimate entities quite incapable of any decision or action whatsoever, selfish or unselfish, as philosopher Mary Midgley and others have pointed out,[31] some going so far as to accuse him of animism.

The Charm of Darwinism; Darwinism as Charm

GIVEN SUCH huge hurdles to credibility, what explanation can there then be for the progress of Darwin's theory through the academy, albeit slow and occasionally discontinuous? Many have scratched their heads over that one. In the somewhat jaundiced opinion of one of Darwin's contemporaries, the then-Keeper of the British Museum's zoological collection, John Grey, it represented a regrettable inconsistency that, while Lamarck had been scorned in Britain for decades, Darwin came along with a theory at least reminiscent of Lamarck's and people such as Lyell and others gave him a respectful hearing.

Was part of the reason for Darwin's success his ability to pull strings in the upper-Middle Class old boys' network of Victorian England? He undoubtedly used his status as a Fellow of the Royal, Linnaean, and Geological Societies to cultivate friendships with influential members of the scientific establishment, and his obsequious tone in letters to members certainly seems to have opened up some doors to him within the scientific establishment. There was also a degree of behind-the-scenes politicking carried out by Darwin and his allies to secure a favorable reception for the *Origin*.[32]

Yet even granting that Victorian Britain was toxically class-ridden, I find it difficult to accept the idea that Darwin's name-recognition and his belonging to the reigning in-group was the reason his theory was eventually accepted. A better explanation may be suggested by the reception of that slightly earlier book which provides a good thematic match for the *Origin*, namely Robert Chambers's *Vestiges of the Natural History of Creation* (1844), discussed in Chapter 3 above. Sir Charles Lyell held that the reason for the success of *Vestiges* lay in good part with the fact of "any theory being preferred to... a series of miracles, a perpetual intervention of the First Cause" (i.e., God).[33] *Mutatis mutandis*, the reason many welcomed Darwin's *Origin* may be found in the same *zeitgeist* of nineteenth-century thinking associated with the positivist philosophy of Auguste Comte, which was to find expression in William Draper's conflict thesis of science and religion.[34]

Comte's opinion that "all real science stands in radical and necessary opposition to all theology"[35] was subsequently translated into a scientific badge of honor by such scientists as Dawkins's avowed role model, nineteenth-century German evolutionary biologist August Weismann, with his insistence that only materialist explanations could be countenanced and all else no-platformed.

It appears that the ideological necessity of finding a strictly materialist theory eventually came to trump those honest and open-minded objections voiced by the majority of reviewers of the *Origin* in the decade after its publication. Intellectual integrity was sacrificed on the altar of ideological commitment. Natural selection was thrust forward aggressively like a form of profane crucifix to ward off the danger thought to be posed by religion, in a way functionally not dissimilar to Voltaire's oft-repeated rallying cry of "écrasez l'infâme" (= "crush the infamous one," by which Voltaire meant the superstitions of religion). Darwinism was beginning to assume for some the function of an anti-religious apotropaic or magic charm against supposed evil.

But as we will see in the next and final chapter, for an increasing number of investigators, the charm is wearing off.

6. Paradigms Regained

I believe that one day the Darwinian myth will be ranked the greatest deceit in the history of science.
—Embryologist Søren Løvtrup[1]

Acceptance of Darwinism has never been universal in the way that, say, acceptance of Einstein's general theory of relativity has. The conceptually flawed and all-too-assailable status of its modern instantiation, neo-Darwinism, has left the door ajar for alternative explanations. The conflict among these competing models is quite complex, with shifting alliances and oppositions depending upon the particular turf of a given skirmish on that great sprawling intellectual battlefield that is origins biology. But the competition, for all this, could be said to finally come down to a simple binary: either matter and energy, wholly in and of themselves, are responsible for all the plants and animals of the world, or else they function together as an instrumentality for some directive agency working through nature. The version of this latter option that is less radioactive to the origins biology community is some updated form of Bergsonian explanation, the idea that the entire process of biological evolution, from the origin of life to the emergence of humanity, was directed from the start by some as yet mysterious internal force, inherent in the cosmic script *ab initio*.

Bergson's invisible force or *élan vital* may seem vague, yet as he noted, we have the precedent of gravity as a force which is indubitably *there* but which we can know about only by inference. Its strength cannot be doubted even though its mainsprings and precise point of origin are unknown, as Newton himself conceded. By the same token, we can plainly observe the way an embryo grows to child and then adult stature and, since we know that this growth and maturation process cannot work by magic or a random jostling of molecules, we can infer that the growth is in response to internal imperatives imprinted within the developing infant *ab ovo*. It is these embedded biological imperatives which accompany the person throughout life's proverbial seven ages, ceasing their agency only at the point of the person's death.

The forms of evolutionary thinking that might be said to be part of this Bergsonian family often fly under the term self-organizational or structuralist. There are those in this camp who would draw a sharp line between what they advocate and what they understand to be advocated by proponents of intelligent design. But what they share in common is worth underlining. Just as Bergson's case for the *élan vital* is circumstantial and inferential, so too is the case for a cosmic designer. So, too, for that matter, is the case for the common descent of all life via random mutations and natural selection. A theory's being circumstantial and inferential can hardly be considered a fatal weakness of any of these arguments, since such is generally the case in origins science. The field, after all, is dedicated to identifying the causes of past events, events that cannot be directly observed. In the case of certain intelligent design arguments, the evidence is numerous and impressive enough to make the inference at least theoretically defensible.

Moreover, if we take the structuralist position and the design position on one side, and set them over against a thoroughgoing evolutionary materialism, the situation becomes clearer still, for without some form of inner or outer directive agency, inert matter—which by definition is without cognitive powers and therefore devoid of any sense of volition—could have had no inclination to create anything at all, even given eons of time. There must therefore be a cause external or internal to matter itself which is responsible for directing matter. Some investigators stop at the invocation of something akin to Bergson's *élan vital*. Others invoke a personal agency, as in the case of the discoverer of quasars, Allan Sandage, who stated, "I find it quite improbable that such order came out of chaos. There has to be some organizing principle. God to me is a mystery but is the explanation for the miracle of existence—why there is something rather than nothing."[2]

Or to frame matters a bit less existentially, we should not be here, and yet here we are. Planet Earth and its welcoming biosphere is a grand cosmic anomaly. That appears to be the common view among many of those best placed to understand what little we can know about the world's origins and evolution. What is the cause of this singular anomaly?

Cosmologist Sir Harold Jeffreys wrote, "I think that all suggested accounts of the origin of the solar system are subject to serious objections. The conclusion in the present state of the subject would be that the system cannot exist."[3] That a world expert in a Cambridge University Press book can say that he is simply flummoxed has the incidental advantage of making me feel somewhat less inadequate about my inability to comprehend how *any* agency could have been responsible for the ultra-complexity of human and animal life.

In addressing these matters, one must balance the paradoxical opinion that we *should not* be here with the certain knowledge that we *are* here. At this point one gains the irresistible impression of science straining at the end of a leash, its reach in sniffing out our place in the cosmos inadequate to the scale of the challenge before it.

I feel even more challenged now than I did before researching this subject, simply because I can now make a more informed estimate of the truly intimidating impenetrability of the enigma to be resolved.

The seal was set on the possibility of any easy solution to the mystery for me by the fact that dialogues on the subject have been so vigorous for the better part of the last three centuries, right up to the present, and conducted by persons on both sides whose intellectual firepower dwarfs my own. In taking in this fact, the impression arises that, if so many persons of such luminous intelligence and ingenuity have bent their minds to solving the problem, and have come up with only the most questionable of hypotheses, then perhaps here is a mystery that will never be wholly unraveled.

When the naturalistic methods of science fail, several possible ways forward present themselves. The first (at times referred to somewhat archly as the "promissory" option) is to express the pious hope that at some time in the future its methods will succeed. This is the unavailing route taken by Darwin with his hoped-for fossil confirmations and by Richard Dawkins waiting for his abiogenetic miracle from the laboratory equivalent to Darwin's small warm pond. The trouble with that option (even setting aside the aforementioned failures) is that it appeals to the old canard that time and opportunity will solve all things, a faith that I, in company with many others, profoundly disbelieve for reasons indicated previously.

A more fruitful option would be to say, "Surely *any* theory must be better than this? Where can we look next?" This would be in essence to make the plea for a Kuhnian paradigm shift—which I take to be the position of James Le Fanu when he concludes that "there must be some prodigious biological phenomenon, unknown to science, that ensures the heart, lungs, sense organs, and so on are constructed to the very highest specifications of automated efficiency."[4]

That for me represents one of the most sensible suggestions I have encountered in a debate more generally marred by appeals to the most hair-raising improbabilities. Of course, it leaves unexamined the precise modalities which might underpin the posited "prodigious biological phenomenon" and does not address the question of how the phenomenon might have been triggered (Le Fanu rejects divine causation). It was observed above that, strictly speaking, attributing creatorial powers to the laws (= regularities) of nature was logically incomplete without asking the further question to make the proposition logically coherent, namely, Who or What was the *legislator* behind these posited laws of nature?

Nevertheless, Le Fanu's idea has the great merit of being a sensibly parsimonious postulate which, being open-ended and essentially interrogative, does not go in for the kind of speculative flights associated with

Darwinian theory. It opens up possibilities for future research without precluding what any results of that future research might turn out to be.

Expanding the Explanatory Toolkit

I DO, however, have one personal caveat about Le Fanu's idea that future researchers should seek the missing biological clues. This is, that it might appear more encouraging had it not been for the tireless (but fruitless) efforts already expended on that quest over the best part of three centuries. That being said, I do not wish to counsel despair and would like to be proven wrong.

Be that as it may, such new research possibilities are for the future. At the moment we are left with the same old conundrums. There is that gigantic pinball machine of the extra-terrestrial universe where everything seems to be a random interplay of blind and devastatingly destructive forces, an impression confirmed by what is coming to be known as the Great Silence, meaning the lack of extra-terrestrial life beyond our privileged abode and the realization that all hoped-for "signals from outer space" have proved illusory—sometimes comically so, as when one set of suspected unearthly signals turned out to be from an all-too terrestrial microwave oven.

Planet Earth, too, in its origin looks to have been little but the accidental detritus of this churning cycle of mega-destruction; and yet over time, fortunately, order emerged from the initial cauldron-like state of our cooling planet. The new condition of habitability appears to have been facilitated by that confluence of benign forces which, although they must have their points of origin somewhere within the Earth's destructive surroundings, appear, happily for us, to represent unprecedented exceptions to the cosmic rule of deadly destruction. The cosmological constants, along with certain local conditions of our corner of the Milky Way, appear to have been synchronized to permit our uniquely habitable planet. The result is that, in contrast to the terrifying scenes of seeming chaos and destruction surrounding us on all sides, we enjoy a planet with a rich and, to date, inexhaustible provision of resources for life.

Moreover, our terrestrial exceptionalism in terms of climate and natural resources alone are just preconditions, inadequate on their own to account for the presence of us, or indeed of even the simplest and first life on planet Earth. So how did life on Earth first arise, assuming that it came about as a result of that chemical fluke referred to as "spontaneous generation?" The abiogenetic nostrum of water + chemicals = organic life has so clearly turned out to be a false hope (except to never-say-die hard-liners and a media commentariat chasing clicks and ratings). Life, it appears, is not just an emergent property of chemistry, so the answers as to how life first arrived to reap the benefits of Earth's bio-friendly conditions must necessarily be sought elsewhere.

The possibility of a material explanation for the mystery confronting us therefore seems to be so vanishingly small that we might simply have to conclude that biogenesis was a one-off, quasi-miraculous occurrence of unknown etiology. The present scientific failure to account for it leaves us precious little alternative than to revisit the possibility of an intelligent mode of causation, even perhaps a supranatural intelligence. That option, which we observed philosopher Antony Flew choosing in a previous chapter, seems to be the only conclusion which is unassailable on strictly logical grounds, however unwelcome that conclusion will seem to many readers, in whose number I count myself. However, it is the only conclusion which I find to be defensible as a logical inference from the data available, a conclusion I arrived at only after assessing the gross explanatory inadequacies of all other theories.

It is in the wake of that comparative exercise that I find myself forced back onto this conclusion, albeit one I endeavor to hold to provisionally and with due modesty, given the scant physical evidence available and the remote nature of the events in question.

I am of course not blind to the paradox of a rationalist and humanist like myself making this argument, but I would plead on my own behalf that it is a logic-driven inference (rather than a philosophic inconsistency) since it owes nothing either to mystical intuition or to the special revelation said to be vouchsafed to us in the Bible or in other foundational books of the world's major religions. It would, as I see the matter, be a betrayal of my rationalist convictions not to follow where the evidence leads merely in order to burnish my credentials among the ranks of more doctrinaire rationalists.

In fact, I would make the suggestion that some of their number may need, like myself, to re-examine what the term rationality might truly consist in today, with special reference to advances made in molecular biology and cosmology in the last half century. As agnostic astrophysicist Paul Davies observed, when we finally come to review an extended explanatory chain, "sooner or later we will have to accept something as given, whether it is God, or logic or a set of laws or some other foundation for existence… whether we call this deeper level of explanation God or something else is essentially a semantic matter."[5]

As Flew's case shows, it is a rich irony of recent history that science, once thought to be the cause of religion's demise, has revealed unsuspected worlds of microscopic precision engineering that goes far beyond the reach of human competence, or even of human comprehension. What sub-Lilliputian equivalents of the Fates of Greek mythology (one might fancifully ask) wove together our molecular destinies? Biological science, once hailed, with more than a little triumphalist glee, as the universal solvent of metaphysical beliefs, is now precisely the force which is making many reassess their philosophical materialism. Time and time again above, in simply following the evidence in the direction in which I judged it led, I have been obliged by the overwhelming force of simple logic to disregard sundry shibboleths of mainstream science as well as my own prior assumptions.

It is not, however, only the discovery of a previously unsuspected microworld and its mind-baffling complexity which has triggered my rethink. A subsidiary reason is my having been alienated by no few scientists' willingness to skew evidence by abjuring the cardinal principle on which I had assumed all science must rest, namely, the absolute, unnegotiable need to provide unbiased evidence for any claims made. To ride roughshod over such a principle means that an idea announced as being scientific under questionable auspices is in reality valueless and can no more lay claim to truth-status than the unsubstantiated speculations of those ancient Greek and Roman natural philosophers considered in Chapter 2. Attempts by some to employ intellectually disingenuous theories to convince us that all life has a natural and discoverable explanation has, in my own case, backfired.

Post-Darwin

THE TERM "scientific conversion" used by Flew has not yet been properly lexicalized. (At the moment I write this, Google refers you only to sundry mathematical conversion tables.) But my hunch is that for the next decade or so it will become more talked about. Moreover, if I have any prediction to offer, it would be that a greater number of people at some not-too-distant future time will see through and past the Darwinian paradigm and so find themselves prepared to start again with a fresh set of questions for biological research.

I am not in principle a great fan of predictions and counterfactual musings, since they depend on just the kind of speculation I have deprecated throughout this volume, and also because history regularly turns out to be far denser with unseen twists and turns than anybody anticipated by extrapolating from prior data. However, we are creatures of foresight, and, if undertaken humbly, there is surely some value in considering future possibilities, and pondering counterfactuals. What if Darwinism is routed? Or to come at the question more imaginatively, what if Darwinism had never won the field? Peter Bowler asked a similar question and hazarded an answer: "Without Darwin's revolutionary input, evolutionism would have developed in a much less confrontational manner, preserving some aspects of the traditional vision of a purposefully designed world and adapting that vision to the modern world via the idea of progress and directed (rather than random) variation."[6]

For instance, had Darwin never lived, it might well have been Alfred Russel Wallace's later, more considered vision of evolution that won the day, one that conceded the limits of natural selection and reopened the door to teleology in biology. Or perhaps it would have been Thomas Huxley's own conceptions which achieved the greater influence, especially since the other competition, Lamarckism, could not (and did not) survive the unarguable counter-indications of Mendelian genetics.[7]

Huxley thought there were "laws of form" determining how structures develop in an organism, without reference to the demands of its environment, a theory called "orthogenesis." On this view, species possessed characteristics unrelated to the demands of adaptation which could not have been formed by natural selection. Purely internal, biological forces propelled them along predetermined paths, and those deeper structures were what permitted the various species to be classified in discrete groups or types by typologists such as, pre-eminently, Richard Owen. Such might have formed the conduit for Owenite ideas to have exerted a greater influence in a way which would have dovetailed with the approach advocated by Michael Denton and other Darwinian naysayers in our own day.

Best of a Bad Lot?

OF COURSE, Darwin cannot and should not be airbrushed out of history. The imaginative exercise above is meant only to broaden our framework of possibilities. The new point of departure suggested by the exercise would, in my view, have a greater chance of making advances than would adherence to a paradigm based on Darwinian assumptions, whose adherents too often appear more intent on hushing up discoveries as to the limits of natural selection than fairly considering alternative hypotheses.

Quite how seriously this modus operandi guides and constricts guild members was exemplified in an incident involving the late Stephen J. Gould of Harvard, that insider's insider to the world of evolutionary science. The incident: he irreverently described "the extreme rarity of transitional forms in the fossil record" as "the trade secret of paleontology,"[8] and the enraged reactions of colleagues quickly made it clear that, in the embattled citadel of Darwinism, no such humor could be permitted. It's hard to blame them: Gould's cavalier disclosure had effectively holed their vessel beneath the water line.

However, the ship has managed to sail on, for if any theory can lay claim to Teflon status (if I may be permitted to vary the metaphor) it is that of natural selection. One example of the resilience of Darwin's ideas even in the face of a direct frontal attack can be found in the mid-1960s, after a group of mathematicians became so disturbed by the optimism of evolutionists about what could be achieved by chance mutations and natural selection that their objections provoked a conference at the Philadelphia Wistar Institute, chaired by Nobel Prize winner Sir Peter Medawar. The conference was entitled Mathematical Challenges to the Theory of Evolution, and in a plenary paper, Professor Murray Eden of MIT put forward probability calculations to argue that neo-Darwinism is mathematically implausible:

> Aside from the pre-Darwinian postulate that offspring resemble their parents, only one major tenet of neo-Darwinian evolution can be said to retain empirical content; namely, that offspring vary from parental types in a random way. It is our contention that if "random" is given a serious and crucial interpretation from a probabilistic point of view, the randomness postulate is highly implausible and that an adequate scientific theory of evolution must await the discovery and elucidation of new natural laws—physical, physico-chemical and biological.[9]

When challenged as to why they still hold to a theory discredited by such mathematical calculations, many to this day will reply with the statement that neo-Darwinism is the "best available" theory to date, so its deficiencies must be tolerated. Setting aside for the moment the question of whether it really is the best available theory, the Darwinists who respond thus seem not to understand that such talk riles members of the general public as well as scholars in other disciplines because it relegates vital issues to the trivial status of a dons' parlor game governed by its own in-house rules and unaccountable to any wider constituency of persons, as Harvard-trained lawyer Norman Macbeth observed a half century ago when he wrote:

I have been rather surprised to discover that many biologists dispute the propriety of a purely skeptical position. They assert that the skeptic is obligated to provide a better theory than the one he attacks. I cannot take this view seriously. If a theory conflicts with the facts or with reason, it is entitled to no respect. As T. H. Huxley long ago remarked, "There is not a single belief that it is not a bounden duty with them [scientists] to hold with a light hand and to part with cheerfully, the moment it is proved to be contrary to any fact, great or small." Whether a better theory is offered is irrelevant.[10]

By the same token, it has been objected that the theory of intelligent design "makes no novel predictions beyond the failure of evolutionary science to explain phenomena"[11] and is in that sense negative rather than constructive. That charge is contested by design theorists, but even if one grants the charge for the sake of argument, should that by itself disqualify the design hypothesis? Surely the important role of intelligent design in the checks-and-balances system of the world of research scientists should be precisely to function in ways analogous to an opposition party in the political arena, in which case it should be allowed to criticize and flag what it views to be the flaws underpinning Darwinian orthodoxy without undue harassment.[12] Honest doubt must be foundational to the scientific method, and no attempt should be made to crush or hobble it on the grounds that it does not tally with the reigning paradigm.

I concur with Norman Macbeth. I find the "best-of-a-bad-lot" defense feeble. I would go further and characterize it as positively misleading. It is rather as if one were to rely on a respected station attendant's directions in finding one's way to a rural train station but, after trekking miles out of one's way and, at the end of the directions, finding oneself in an empty field, refusing to turn back and seek a better route on the grounds that the information received was the "best available." No, the sensible thing is to abandon the empty field and the daft directions.

If not Darwin, who or what? I confess to finding a certain appeal in non-Darwinian theories such as Bergson's, which at least posit some driving force behind a process which otherwise would have no instrumental capability and means of forward propulsion at all. Bergson's ideas have attracted later scientists such as Hans Driesch in the earlier part of the twentieth century and Rupert Sheldrake in the late twentieth century.[13]

But I also have my reservations. Bergson's idea of a vital force, encapsulated by Peter Bowler as "a spiritual force imposing a rational order on the development of life,"[14] was rejected by Wallace, notwithstanding his later theistic turn. The co-discoverer of the theory of evolution by natural selection saw in Bergson's theory "vague ideas" of "no real value as an explanation of Nature."[15] I have to say that for myself, too, Bergson's invocation of an *élan vital* seems hardly more informative than if he had used the expression *je ne sais quoi* (a quality that cannot be described or named easily). For *élan vital* seems to be but a placeholder term for what in reality is only a vague intimation of some unspecified agency. And since the whole exercise is to identify the agency or agencies for the origin of new form in the history of life, what then has been accomplished?

If we are to eschew grand-sounding but ultimately vacuous cop-outs, we may just have to live with the fact that there is an order of reality resistant to human apprehension, and simply resolve to "get over it." Whether we like it or not, we may be fated to remain largely uninitiated spectators of life's unexplained pageant rather than players who understand its genesis and modalities. Facing this possibility, it may be difficult to resist the unwelcome presentiment that we might all have been press-ganged by some prank-playing cosmic joker as bit players in some dismal cosmic drama of the absurd.

To give over wholly to such a feeling would plunge us headlong into the gloomy waters of those existentialist philosophers, novelists, and dramatists—Kierkegaard, Sartre, Camus, Becket, et al.—a subject which, thankfully, is beyond my present remit and around which I shall place a *cordon sanitaire!*[16] Better, perhaps, to merely dip a toe in such waters and welcome as salutary a measure of epistemological humility sufficient to steer us around the hubristic and psychologically maladaptive response of laying claim to knowledge we do not possess, or sallying forth on some quasi-Faustian quest for the unknowable with all its attendant risks to our psychological well-being.

I don't mean here to negate my earlier confession of finding myself surprised by the relative strength of the design hypothesis in the wake of several modern discoveries in cosmology and biology. Oxford theology professor emeritus Richard Swinburne commented to *The New Statesman* that "to suppose that there is a God explains why there is a physical universe at all; why there are the scientific laws there are; why animals and then human beings have evolved.... In fact, the hypothesis of the existence of God makes sense of the whole of our experience and it does so better than any other explanation that can be put forward, and that is the grounds for believing it to be true."[17] I do indeed find those views compelling. But in an area of enquiry in which there is a notorious dearth of certainties, equal time should be given to the opinion that the Baconian virtue of "nescience" should be practiced, by which is meant not culpable ignorance but rather an acknowledgement that "we do not know." Such a practice would guard us from the temptation of making claims to knowledge we do not possess.

A similar, essentially agnostic idea was recently put forward by American philosopher Thomas Nagel, in the course of his questioning of sundry simplistic or reductionist explanations sailing under the false flag of science. His conclusion is blunt: We are no more able to understand what Lucretius called the nature of things than could Aristotle—a contention borne out when we look at the insoluble crux of the human brain and the subject of consciousness. The brain's fantastical degree of micro-organization and its ability to communicate with our limbs to produce just the right movements could hardly be accounted for by a blunt process incapable of harmonizing coordinating functions.

In the interest of full disclosure, I have to confess that the human brain seems to me to be so ineffably complex that I sometimes find it difficult to conceive of *any* agency being responsible for its sublime intricacies—*even a divine one*! Indeed, extending this truth-or-dare riff a trifle, I have to make the further confession that I harbor the same bafflement with regard to what possible agency could have been (ultimately) instrumental in the creation of, for instance, our family pets or the butterflies on our lawn. Hence the bathetic notion that natural selection could have played any part in creating the human brain inevitably seems to me to point to that rather steep descent from the sublime to the ridiculous.

I am aware that persons who defend the role of natural selection in the formation of plant and sentient life on planet Earth will "have a laugh" at my confession, but from my perspective it is the bold claims for natural selection that strike me as risible. It is for the reasons noted above that I have to conclude, rather bluntly I fear, that as far as hyperbolic fantasies about solving the mystery of life are concerned, the moral would seem to be to exercise a little more humility in the face of life's perennial mysteries and acknowledge a real if sobering possibility: that we may in all this be faced with, to borrow the Chomskyan distinction again, a mystery and not a (soluble) problem. Or, as another distinguished scientist was once moved to concede, "By the very nature of things no one can know with absolute certainty how living things arose in past ages."[18]

That conclusion, from one of the leading experts in the middle of the twentieth century, is no less valid today. As Paul Davies commented more recently, we can pursue rational enquiry till the cows come home, but "my instinctive belief [is] that it is probably impossible for poor old *homo sapiens* to get to the bottom of it all."[19]

Prometheus Unhinged

THE PRIMARY and most significant existential challenge we face in our lives, it appears to me, is to stake out a delimited arena of meaning for ourselves and contrive to seek fulfilment in that "much in a little compass." Otherwise we run the risk of being ruled by the rock rather than the rudder. Darwin found this out to his cost in the over two decades of torment he experienced from the time of the publication of *Origin* up to his death in 1882. When he died, he had not yet advanced to the posthumous status of the unassailable Sage of Down bestowed on him by many modern quasi-hagiographical historians of science. Hence St. George

Mivart and others didn't hesitate to impress upon him the many "incon-sistencies and ambiguities"[20] in his theory, and he went to his grave with them still weighing heavily upon his shoulders. Not for nothing has A. N. Wilson compared Darwin in older age with Hamlet.

Mivart's unremitting attacks on Darwin's theory undoubtedly played a role in Darwin's decision to abandon the arena of combat. Dar-win had little to say publicly on the subject, but his two private written remarks, both made public only after his death, are instructive. In an 1860 letter to Asa Gray, he laid bare his doubts and confusion:

> There seems to me too much misery in the world. I cannot persuade myself that a beneficent & omnipotent God would have designedly created the Ichneumonidæ with the express intention of their feed-ing within the living bodies of caterpillars, or that a cat should play with mice. Not believing this, I see no necessity in the belief that the eye was expressly designed. On the other hand I cannot anyhow be contented to view this wonderful universe & especially the nature of man, & to conclude that everything is the result of brute force. I am inclined to look at everything as resulting from designed laws, with the details, whether good or bad, left to the working out of what we may call chance. Not that this notion at all satisfies me. I feel most deeply that the whole subject is too profound for the human intellect. A dog might as well speculate on the mind of Newton.[21]

In his autobiography, written in 1878 and published only after his death, he again called attention to the problem of animal suffering as weighing against the idea of a benevolent God, but then, two pages later, and as already noted above, he comments that when reflecting upon "this immense and wonderful universe, including man with his capacity of looking far backwards and far into futurity," he feels "compelled to look to a First Cause having an intelligent mind in some degree analogous to that of man; and I deserve to be called a Theist." This is then followed by further wrestling and doubting, after which he concludes that "I for one must be content to remain an Agnostic."[22]

Even this label he wore tentatively, emphasizing that his "judgment often fluctuates." As he explained in an 1879 letter to John Fordyce,

"Whether a man deserves to be called a theist depends on the definition of the term... In my most extreme fluctuations I have never been an atheist in the sense of denying the existence of a God. —I think that generally (and more and more so as I grow older), but not always,—that an agnostic would be the most correct description of my state of mind."[23]

Such thoughts indicate that even in later years he was never able to satisfactorily resolve the conflict in his mind between a naturalistic and a theistic (or rather, deistic) understanding of the nature of things (a conflict shared by no few scientists of the modern era). He was sometimes even beset by doubts that his life's work had been based on an ill-conceived fantasy. There is a somewhat disturbing parallel between Darwin in older age and Mary Shelley's figure of Victor Frankenstein, the "Modern Prometheus" of her subtitle. Whereas the politically radical Romantic revolutionary Percy Bysshe Shelley shows an understandable sympathy for that iconic representative of hubris in his lyrical drama, *Prometheus Unbound*, his wife's novel shows the negative consequences of her hero's tampering with the mysteries of creation in a way which reinforces the moral implication of the myth as it found expression in classical antiquity.

It will be recalled that when Prometheus stole fire from the Greek gods, Zeus punished him, *inter alia*, by creating the infamous Pandora who unleashed the evils of hard work and disease on humanity when she removed the lid of her famous box (or jar). Mary Shelley, as her subtitle makes clear, transfers the spirit of the old mythology to her own day. Her novel is often interpreted, I think rightly, as having been in part a literary riposte to the same overreaching masculinist ethos (a.k.a. hubris) displayed both by her husband and by their scientist friend, Sir Humphry Davy, who once delivered himself of this vaunting estimate of the boundless powers of science:

> Science has bestowed upon mankind powers which may be called almost creative which have enabled him to change and modify the beings surrounding him, and by his experiments to interrogate nature with power, not simply as a scholar, passive and seeking only to understand

her operations, but rather as a master, active with his own instruments. Who would not be ambitious of becoming acquainted with the most profound secrets of nature; of ascertaining her hidden operations and of exhibiting to man that system of knowledge which relates so intimately to their own physical and moral constitution?[24]

Frankenstein is in good part a warning against such overweening scientific arrogance.

By contrast with Darwin, Wallace was never compelled to face the older man's torments, because he had resolved the same tension in favor of the theistic alternative; and although he was inevitably the recipient of much scientific opprobrium for his tergiversation, there can be little doubt that, with his mind made up and his tensions banished, his older age was considerably happier and more productive than that of his peer. He became a respected lecturer, published widely on a broad range of topics, had a fulfilled family life, and was even able to afford a substantial house just south of London which he could have only dreamed of in his impecunious youth, when his father was forced to move the family from London to Usk in South Wales to seek relief from the heavy cost of metropolitan living.

To be sure, practicing the self-denying ordinance of curbing one's appetite for pat solutions is easier said than done. Wallace proved himself able, eventually retracting his grand claims for natural selection, since the theory accorded so ill with the facts as he later saw them. Darwin spent the rest of his life wracked by doubts about his own theory, doubts augmented by the criticisms of early reviewers whose objections he felt honor-bound to integrate into later versions of his eventually much-diluted treatise.

The mystery of biological origins remains. Concerning the origin of the first life, recent contributors to a *New Scientist* guide for young people gamely reference the old chestnuts of clay forming a prebiotic substrate at the bottom of a pond, thermal/volcanic vents, and (in a heavily caveated reference) panspermia, but finally, I was relieved to read, they make the honest disclosure that "we may never uncover the answer."[25] It is past due that such a frank admission was extended to the origin of biological diversity after the first life, to Darwin's theory of evolution and its various evolutionary descendants. The elephant of the creation/evolution enigma sits as immovably in our drawing rooms as it did in 1858 and in the decades, centuries, and millennia prior to that.

In this remarkable historical about-turn, the wholly unanticipated philosophic, even potentially theistic developments brought about by the contemporary intelligent design movement's critique of modern evolutionary theory gives a form of belated vindication to two of Darwin's contemporaries thought to have been consigned to historical footnotes, namely, Harvard professor Louis Agassiz and Darwin's Cambridge tutor, Adam Sedgwick. Both men thought that science's inability to plumb the mysteries of nature was itself evidence of the divine. As historian Neal Gillespie observed, "There was in some—one thinks of Sedgwick and Agassiz—a pious gladness in this inability to probe to the depths of the secrets of nature, as if God's being was glorified in man's weakness."[26] Both men, if I might be permitted to indulge my imagination for a moment, could very well (posthumously!) be enjoying the last laugh.

The Materialist Paradigm: A Flawed Hypothesis?

ALTHOUGH THE Earth is thought to have emerged from the same material chaos as that which presently characterizes much of the rest of the observable cosmos, its mysterious acquisition of a biosphere benignly pullulating with all sorts and conditions of life-forms has meant that our planet is quite unusual, perhaps unique, in having been able to transcend its lifeless origins. The numerous ways it appears fine tuned for life give us pause. So too does the emergence of the first life. Nobel Prize holder Harold Urey once issued the challenging brain-teaser that the more he and others study the origin-of-life question, the more it appears that life "is too complex to have evolved anywhere."[27]

Conclusions adducing a wholly natural causation of the biosphere, on the other hand, tend to be skewed by an undiscriminating lumping together of the Earth with all those regions of the cosmos beyond

Earth. Such an analogy is misleading since, as of yet, we have found only what is lifeless beyond our world, and the gulf between life and non-life is unimaginably great. Our living planet is so unusual as to require an explanation over and above what is required to explain the evolution of galaxies, stars, and lifeless planets. Indeed, it is difficult *not* to deduce that we must be dependent on a special dispensation, for no alternative logical pathway presents itself. To insist that such an alternative is verboten would seem to betray a form of ideological resistance and denial which, I suspect, lurks behind many attempts to postulate a purely material genesis for our biosphere.

Such attempts stretch back to encompass efforts to explain our finely tuned planet in strictly materialistic terms. It extends to the tall-tale claims for the spontaneous generation of first life. And it extends beyond the origin of life into the obsessive shoehorning of all available evidence into a Procrustean, quasi-Malthusian schema of natural selection, even if in some quarters augmented by ancillary evolutionary hypotheses such as genetic drift, neutral evolution, and other purely a-teleological add-ons, continuing in the plucky tradition of Darwin's later editions of the *Origin*. Such efforts have been responsible for those many disingenuous reasonings, cognitive dissonances and, frankly, credulities which have afflicted biologists from the time of Darwin down to our own day.

Far better and more credible in my view is to stop digging the hole than to continue excavating and defending implausible theories in ways not only offensive to the empirical principles of good science but also potentially obstructive to any future research conducted under the aegis of a modified or contrary paradigm. As Robert Shapiro commented, "Some theories come labeled as The Answer. As such they are more properly classified as mythology or religion than as science."[28] The point of such bluffing is to inoculate "The Answer" from free and fair competition with competing explanations.

Rigid adherence to such theories in defiance of reasonably presented counter-indications also runs the risk of squandering public trust. In the interest of keeping faith with members of the public outside the professional science guild, it might be preferable for biological specialists to come clean about their ignorance of, or even ambivalence toward, baffling phenomena, which anyone who wishes to grapple with our place in a bewildering universe must confront. Having to make profound existential choices about the values and beliefs we choose on the basis of half-truths and studied obfuscations clearly can do the general public no good. A best practice based on the principle of straight candor is surely indicated.

If the reigning materialist paradigm had even a tolerably convincing weight of evidence behind it, I would be the first to accept it. In fact, I would embrace it wholeheartedly and with a sense of relief, even closure, since it would provide an excellent fit with a prior educational formation which has habitually foregrounded rational, evidence-based criteria. However, it is those very rationalist principles which bid me reject the Darwinian narrative, in its original, neo-Darwinian, and extended manifestations. I find it the grandest historical irony that the most fervent defenders of Darwinism claim to be advancing the ideals of the European Enlightenment. My view is that they are in reality dishonoring the foundational principles of that admirable project by perpetuating a hypothesis without empirical foundation or even the slightest approximation to verisimilitude.

As philosopher Richard Spilsbury once noted, "The basic objection to neo-Darwinism is not that it is speculative, but that it confers miraculous powers on inappropriate agents. In essence, it is an attempt to supernaturalize nature, to endow unthinking processes with more-than-human powers."[29]

The case might even be made that the Darwinian narrative can work only by implicitly disregarding the Enlightenment program through its appeal to ways of thought supposed to have died out countless centuries before Darwin was even born. By that I mean that to attribute creative potential to nature itself is a deeply archaic, animistic way of thinking which takes us back even to the paganism of the Homeric age.

In the imaginative works of those early eras, nature through its many deified incarnations is routinely credited with directive capability. Zeus, called The Thunderer by the poet Hesiod in his *Theogony*, was believed to be able, *inter alia*, to control the weather; Demeter, the fertility goddess, could exert an influence on the annual crop yield; Aeolus, Keeper of the Winds in the *Odyssey*, provides a gentle breeze to waft Odysseus back to Ithaca after his long travels. To the ancient Greeks and many peoples who preceded them, the gods were essentially personifications of different aspects of Nature itself. The pre-scientific mind imputed agency to Nature by way of the personification of Nature's various aspects as individual divinities.

Darwin's theory of natural selection, although it struck most at the time and even since as an intellectual innovation, appears in reality to be something of a throw-back to those earlier modes of thought. In what seems to be a confirmation of the "nothing new under the sun" adage, Darwin appears, wittingly or not, to have channeled the spirit of the older, polytheistic world by crediting Nature with an infinite number of transformative powers. This particular objection to Darwinism has not been specifically adverted to in the numerous publications I have sampled, though as noted, Wallace and Bishop Wilberforce did object to the extravagant powers Darwin granted to natural selection.

The equation of nature with divine forces was a phenomenon well understood by pre-scientific communities, but it is a mental world which we supposedly lost many centuries ago and, on the face of it, it would carry little credibility today to impute human-like agency to any aspect of external nature. That Shakespeare still retained a feeling for such thought-ways is evident in his deployment of the dramatic technique of pathetic fallacy. Even in the England of circa 1600, however, I suspect that the supremely versatile dramatist may have been giving poetic expression to an obsolescent belief. Yet whatever the precise phenomenological status of Nature might have been in the minds of our Elizabethan forbears, there can be little dispute that in the third decade of the twenty-first century, despite our ready ability to warm to Shakespearean pathetic fallacy as a marvelous poetic device, we no longer understand it literally and viscerally as a logic relevant to our own lives (unless we are of a particularly mystical bent).

If, then, we no longer believe nature to possess power in the way familiar to many ancient Greeks and Mesopotamians, natural selection can only appear as an outmoded postulate void of instrumental capability. Since nature is no longer thought to contain the directive force of any immanent divinity, it might now seem (picking up on my earlier analogy) to be functionally as powerless as an inert metal chassis without an engine. To claim that such an unpowered vehicle, so far from being doomed to everlasting stasis, could have been the driver of all those vast transmutations responsible for populating the earth in all its diverse profusion should strike us as unconvincing. In fact, the idea might appear incomprehensible by the light of those rational criteria used by citizens of the post-Enlightenment age to understand the world.

The attempt to solve the mystery of speciation by positing a selection procedure initiated and implemented by unaided nature falls at every hurdle. It lacks explanatory force, empirical foundation, and logical coherence. It postulates the contradiction-in-terms of a metamorphosizing, species-creating dynamic issuing from a process lacking any discernible dynamic. It is ultimately a pseudo-explanation, a way of concealing underlying ignorance.

So unconvincing must this archaic thought-pattern seem to the modern, scientifically literate mind (one would have thought!) that, once recognized for what it is, its unintended consequence can only be to reinforce the alternative position of divine causation. This, of course, is precisely the option rejected by countless atheists and agnostics in the post-Enlightenment centuries, with the result that such persons must necessarily find themselves stranded between the devil and the deep blue sea when faced with their unenviable choice.

However sympathetic I am to that plight, as one who has occupied precisely that uncomfortable space, I find that the most rational conclusion of this rationalist is to view as the default position the hypothesis that sentient life could not have developed without some form of fore-sight and an accompanying instrumental power to realize that vision in practice. This must point us away from nature itself as the sufficient causal power for that "mystery of mysteries" and in the direction of an unknown (and potentially unknowable) source of intelligence outside nature—and, to judge from the sublime intricacies and spellbinding wonders with which our world abounds, a supra-human form of intel-ligence at that.

The genesis and evolution of our fine-tuned cosmos and biosphere must in the end come down to a clear binary: either nature did the fine tuning and selecting or God did (however that latter entity may be con-ceived and glossed). To say that "God did it" obviously does not sit well with people holding a non-theistic worldview. To say that "nature did it" arguably carries even less plausibility, so that many persons may feel themselves torn between two equally improbable positions. However, with the naturalistic/materialistic alternative having failed so signally, we are left with no other choice but to consider the possibility of the "God hypothesis." Faced with the sheer unfeasibility of a purely natu-ral explanation, logic leaves us with little other choice. Extending the old adage that nothing comes of nothing, it might be contended that in real life, in contradistinction to the magician's claim of a rabbit magically emerging from the hat, nothing can "magically emerge" or "naturally evolve" without a supporting agency—little though we may know of that originating agency. In default of a better explanation than that offered by the Darwinian paradigm and its various materialistic descendants and kissing cousins, however, this hypothesis surely cannot be discounted out of hand.

EPILOGUE

WHEN MY WIFE AND I VISIT RURAL BRITTANY, ONE OF OUR FA-vorite ports of call is a lovely coastal church called St. Jean du Doigt (Saint John of the Finger), where the eponymous digit of the apostle is popularly supposed to be stored. For us this quaint belief adds to the unspoiled charm of the Breton countryside. Historically the medieval practice of collecting saints' relics is now of course commonly regarded as a form of "pious fraud," a means of buttressing the power and influence of the Catholic Church in the Middle Ages. What struck me recently is that the instrumentalization of an unverifiable, non-evidence-based hypothesis to prop up today's secular ideology presents a telling mirror image of the medieval practice. Given the secularizing *volte-face* experienced in post-Christian Europe, an important motive for giving such an easy pass to the quasi-magical notion of natural selection seems to be the desire to deter people from entertaining any notion of divine creation.

If anything, this modern form of hoodwinking seems less forgivable than its medieval variant, since it is so out of line with the values of our "age of the masses" (to borrow the title of Michael Biddiss's classic), an age of universal suffrage and democracy where each individual has the right to make up his or her mind. To allow and abet a deception to be practiced upon people in the attempt to prevent them making up their own minds about something as fundamental as their preferred existential position in life is to my mind as misguided and paternalistic a practice as any perpetrated by the medieval Church.

It is now half a human lifetime since Michael Denton issued his decisive critique of modern evolutionary theory, and yet many biologists continue on, business as usual. School textbooks still purvey the same broadly Darwinian interpretation of life, including the presentation of "evidences" for evolution that have long since been discredited.[1] Richard Dawkins was recently given a very easy ride by Mark Urban on BBC's Newsnight program.[2] And in the teeth of all empirical evidence to the contrary, contemporary Darwinism has become accepted as the most grown-up form of understanding of humankind's existential status by the many who, I suspect, have had little time or opportunity to "fact check" the propositions they are buying into. For more and more people, this acceptance seems to be a wholly unexamined assumption.

However, not all have been so unenquiring or supine, and Richard Dawkins has even been stung to lament the fact that outsiders have presumed to question the assumptions of biology specialists whereas they disregard what goes on in other branches of science such as, say, quantum theory. The reason for this is of course (as he must surely know) that his particular discipline holds such vast implications for the existential situation of all men and women, for the very "ground of their being," such that many quite rightly find it impossible to ignore. If nothing else has been achieved in this short volume, I hope, by presenting views which differ from current orthodoxy, to have given readers the chance to reflect with me on the many problematical facets of modern evolutionary theory, and to grapple with the possible implications of evidence that does not easily accord with materialism.

My own position, as a long-standing humanist with no allegiance to any revealed faith, remains that we each have to come to terms with an inscrutable universe in the best, and most morally accountable, way we can. Others should be free to come to their own conclusions on an issue in which there may be no unalloyed truth-bearers, only truth-seekers, in whose number I very much still count myself.

ENDNOTES

PROLOGUE

1. Charles Darwin, *On the Origin of Species by Means of Natural Selection, or the Preservation of Favoured Races in the Struggle for Life* (London: John Murray, 1859), http://darwin-online.org.uk/converted/pdf/1859_Origin_F373.pdf. Darwin dropped the word "on" from the title in the 6th edition.

2. On the sometimes counter-intuitive weirdness of science see Lewis Wolpert, *Six Impossible Things before Breakfast: The Evolutionary Origins of Belief* (London: Faber and Faber, 2006) and *The Unnatural Nature of Science* (London: Faber and Faber, 1993).

3. Charles James Fox Bunbury, letter to Katharine Lyell, October 31, 1859, in *The Life of Sir Charles J. F. Bunbury, Bart.*, ed. Mrs. Henry Lyell (London: John Murray, 1906), 2:149, http://darwin-online.org.uk/content/frameset?viewtype=text&itemID=A716&pageseq=1.

4. James Boswell, *Dr. Johnson's Table-Talk: Containing Aphorisms on Literature, Life, and Manners; with Anecdotes of Distinguished Persons: Selected and Arranged from "Dr. Boswell's Life of Johnson"* (London: C. Dilley, 1798), 394.

5. Noel Annan, *Leslie Stephen: The Godless Victorian* (New York: Random House, 1984).

6. Charles Darwin, *The Descent of Man and Selection in Relation to Sex* [1871], eds. James Moore and Adrian Desmond (London: Penguin, 2004).

7. Psalm 8:5.

1. THE BATTLE IS JOINED

1. Thomas Nagel, *Mind and Cosmos: Why the Materialist Neo-Darwinian Conception of Nature Is Almost Certainly False* (Oxford: Oxford University Press, 2012), 6, 128.

2. F. C. S. Schiller, "Darwinism and Design," *The Contemporary Review* 71 (June 1897): 867–883. The article was later reprinted in a collection and can be read here: https://archive.org/details/cu31924029012171.

3. Thomas Malthus, *An Essay on the Principle of Population* [1798], ed. Antony Flew (London: Penguin, 1970).

4. Thomas Hobbes, *Leviathan* [1651], ed. Christopher Brooke (London: Penguin, 2017), 100–103.

5. Charles Darwin, *The Autobiography of Charles Darwin 1809–1882*, ed. Nora Barlow [1958] (London: Norton, 1993), 120.

6. Alfred Russel Wallace, *My Life* (London: Chapman and Hall, 1905), 2:232.

7. Charles Darwin, *On the Origin of Species by Means of Natural Selection, or the Preservation of Favoured Races in the Struggle for Life* (London: John Murray, 1859), 482, http://darwin-online.org.uk/converted/pdf/1859_Origin_F373.pdf.

8. Darwin, *Origin of Species*, 481.

9. Asa Gray, "Darwin on the Origin of Species," *The Atlantic*, July 1860, https://www.theatlantic.com/magazine/archive/1860/07/darwin-on-the-origin-of-species/304152/.

10. Charles Lyell, *Geological Evidences of the Antiquity of Man with Remarks on Theories of the Origin of Species Variation* [1863] (Cambridge: Cambridge University Press, 2009), 505.

11. Lyell, *Geological Evidences*, 506.

12. This subject is dealt with very fully by James R. Moore, *The Post-Darwinian Controversies: A Study of the Protestant Struggle to Come to Terms with Darwin in Great Britain and America 1870–1900* (Cambridge: Cambridge University Press, 1981).

13. Charles Kingsley to Charles Darwin, November 18, 1859, Darwin Correspondence Project, Letter no. 2534, University of Cambridge, https://www.darwinproject.ac.uk/letter/DCP-LETT–2534.xml.

14. Charles Darwin to Charles Lyell, October 11, 1859, Darwin Correspondence Project, Letter no. 2503, University of Cambridge, https://www.darwinproject.ac.uk/letter/DCP–LETT-2503.xml.

15. David Waltham, *Lucky Planet: Why Earth Is Exceptional—And What That Means for Life in the Universe* (New York: Basic Books, 2014), 109.

16. See on this trend Rudolf Bultmann, *Jesus Christ and Mythology* [1926] (London: SCM, 2012).

17. Richard Dawkins, *The Magic of Reality* (London: Transworld, 2012), 30.

18. Charles Darwin to George Charles Wallace, March 28, 1882, in Gavin De Beer, "Some Unpublished Letters of Charles Darwin," *Notes and Records of the Royal Society of London* 14, no. 1 (June 1959): 59.

19. Charles Darwin to Joseph D. Hooker, February 1, 1871, Darwin Correspondence Project, Letter no. 7471, University of Cambridge, https://www.darwinproject.ac.uk/letter/?docId=letters/DCP-LETT-7471.xml.

2. The Evolution of a Myth

1. William H. Thorpe, *Purpose in a World of Chance: A Biologist's View* (Oxford: Oxford University Press, 1978), 6.

2. Charles Darwin, *The Voyage of Charles Darwin: Autobiographical Writings*, ed. Christopher Rawling (London: BBC, 1979), 19.

3. Adrian Desmond and James Moore, *Darwin: The Life of a Tormented Evolutionist* (London: Penguin Books, 1991), 31.

4. Darwin, *The Voyage of Charles Darwin*, 22.

5. For more on this, see Harold L. Burstyn, "If Darwin Wasn't the *Beagle*'s Naturalist, Why Was He on Board?," *The British Journal of Historical Science* 8, no. 1 (March 1975): 62–69.

6. William Paley, *Natural Theology: Or Evidence of the Existence and Attributes of the Deity, Collected from the Appearances of Nature* [1802], ed. Matthew D. Eddy and David Knight (Oxford: Oxford University Press, 2008).

7. Paley, *Natural Theology*, 7–8.

8. David Hume, *Dialogues Concerning Natural Religion* [1779], Part VII, in Hume, *Principal Writings on Religion Including "Dialogues Concerning Natural Religion" and "The Natural History of Religion"* ed. J. C. A. Gaskin (Oxford: Oxford University Press, 1993), 82.

9. Hume, *Dialogues Concerning Natural Religion*, Part VIII, 88–89.

10. Matthew Arnold, "Dover Beach," *New Poems* (London: MacMillan, 1867), lines 21–28.

11. Alfred, Lord Tennyson, *In Memoriam AHH* (London: Edward Moxon, 1850), canto 55, lines 13–16.

12. Tennyson also was influenced by Robert Chambers's *Vestiges of the Natural History of Creation* (London: John Churchill, 1844).

13. David Knight, *Science and Spirituality: The Volatile Connection* (London: Routledge, 2004), 60.

14. James Hannam, *The Genesis of Science: How the Christian Middle Ages Launched the Scientific Revolution* (Washington, DC: Regnery, 2011), 321. See also Beryl Smalley, *The Study of the Bible in the Middle Ages* (Oxford: Clarendon Press, 1952).

15. The Victoria Institute's archives are housed at the University of Manchester, https://archive-shub.jisc.ac.uk/manchesteruniversity/archives/e5d3f9f2-da61-39c3-a7ba-276c538f3e3a.The Victoria Institute later became Faith and Thought, https://www.faithandthought.org/about. html.

16. Effie Munday, "The British Evolution Protest Movement: A Brief History," *Creation* 8, no. 2 (March 1986), 41–42, https://creation.com/evolution-protest-movement.

17. Douglas Dewar, *Difficulties of the Evolution Theory* (London: Edward Arnold, 1931); *More Difficulties of the Evolution Theory* (London: Thynne and Co., 1938).

18. Alvar Ellegård, *Darwin and the General Reader: The Reception of Darwin's Theory of Evolution in the British Periodical Press 1859–72* (Gothenburg: Elanders, 1958).

19. See Ellegård, *Darwin and the General Reader,* 118–119.

20. *The Eclectic Review* (1859): 557, quoted in Ellegård, *Darwin and the General Reader,* 117–118.

21. Malcolm Muggeridge, *The Earnest Atheist* (London: Eyre and Spottiswood, 1936), 221.

22. Samuel Butler, *Evolution Old and New* [1879] (New York: Jefferson, 2016).

23. For more on the question of Darwin's predecessors and the controversy over whether and to what degree he plagiarized some elements of his theory, see Rebecca Stott, *Darwin's Ghosts: The Secret History of Evolution* (London: Bloomsbury, 2012), and John G. West's critical review of the book, "Dissent of Man," *Claremont Review of Books* 13, no. 2 (Spring 2013), https://claremontreviewofbooks.com/dissent-of-man/.

24. Peter Raby, *Samuel Butler: A Biography* (London: Hogarth, 1991), 169. See also Muggeridge, *The Earnest Atheist,* 218, for Butler's pantheism.

25. William Hale White, *The Autobiography of Mark Rutherford,* ed. William S. Peterson (Oxford: Oxford University Press, 1990). Rutherford, we are told, rekindled his dwindling faith after reading the *Lyrical Ballads.*

26. William Wordsworth, "Lines Composed a Few Miles above Tintern Abbey," *Lyrical Ballads* (London: Arch, 1798), lines 91–100.

27. Loren Eiseley, *Darwin's Century: Evolution and the Men Who Discovered It* (Garden City, New York: Doubleday, 1958), 124.

28. Janet Browne, *Charles Darwin: Voyaging* (London: Pimlico, 2003), 83.

29. Michael Flannery, *Intelligent Evolution: How Alfred Russel Wallace's World of Life Challenged Darwinism* (Riesel, TX: Erasmus Press, 2020), 16–17.

30. Philip G. Forthergill, *Historical Aspects of Organic Evolution* (London: Hollis and Carter, 1952), 13–14. For more on Anaximander and Anaximenes and other pre-Socratic philosophers, see G. S. Kirk and J. E. Raven, *The Pre-Socratic Philosophers* (Cambridge, UK: Cambridge University Press, 1957).

31. Neal C. Gillespie, *Charles Darwin and the Problem of Creation* (Chicago: Chicago University Press, 1979), 105.

32. Stephen Greenblatt, *The Swerve: How the Renaissance Began* (London: Vintage, 2012). For the text of Lucretius in English translation see *The Nature of Things,* trans. A. E. Stallings (London: Penguin, 2012).

33. George K. Strodach, introduction to *The Art of Happiness,* by Epicurus, trans. George K. Strodach (London: Penguin, 2012), 7.

34. David Hume, *Dialogues and Natural History of Religion* [1779], ed. J. C. A. Gaskin (Oxford: Oxford University Press, 2008).

35. Quoted in Maureen McNeil, *Under the Banner of Science, Erasmus Darwin and his Age* (Manchester, Manchester University Press, 1987), 98.

36. Erasmus Darwin, quoted in Charles Coulton Gillispie, *Genesis and Geology: The Impact of Scientific Discoveries upon Religious Beliefs in the Decades before Darwin* (New York: Harper and Row, 1959), 33.

37. Erasmus Darwin, *Zoonomia; Or, the Laws of Organic Life* (London: J. Johnson, 1794), section XXXIX: Of Generation, https://www.gutenberg.org/files/15707/15707-h/15707-h.htm.

38. It is tempting to view Darwin's enviable knack of developing companionable relations with his academic tutors and sharing in their intellectual discussions as purely a case of family pedigree opening doors that might have remained closed to less well-connected students, especially in the context of a then totally unreconstructed English old-boy network in which it was often who rather than what you knew which proved the greater advantage. However, this does not account for Charles's sustained discussions with men who were the very antithesis of intellectual gadflies and who would have been able to see through their charge in an instant had his interest in their subject specialties been feigned or too superficial to merit sustained dialogue.

39. Cited by Janet Browne, *Charles Darwin: Voyaging* (London: Pimlico, 2003), 39.

40. James A. Secord, *Victorian Sensation: The Extraordinary Publication, Reception and Secret Authorship of "Vestiges of the Natural History of Creation"* (Chicago: Chicago University Press, 2000).

41. Robert Chambers, *Vestiges of the Natural History of Creation* (London: John Churchill, 1844), 208.

42. Chambers, *Vestiges*, 235.

43. Charles Darwin to Thomas Henry Huxley, September 2, 1854, Darwin Correspondence Project, Letter no. 1587, University of Cambridge, https://www.darwinproject.ac.uk/letter/DCP-LETT-1587.xml.

44. Charles Darwin to Joseph D. Hooker, February 1, 1871, Darwin Correspondence Project, Letter no. 7471, University of Cambridge, https://www.darwinproject.ac.uk/letter/?docId=letters/DCP-LETT-7471.xml.

45. Richard Dawkins, *The Blind Watchmaker* (London: Penguin, 1986), 43.

46. Matti Leisola and Jonathan Witt, *Heretic: One Scientist's Journey from Darwin to Design* (Seattle: Discovery Institute Press, 2018), 23.

47. "Science: Semi-Creation," *TIME*, May 25, 1953, http://content.time.com/time/subscriber/article/0,33009,890596,00.html.

48. Carl Sagan, quoted in Robert Shapiro, *Origins: A Skeptic's Guide to the Creation of Life on Earth* (New York: Summit Books, 1986), 105.

49. For more on the Miller-Urey experiment and how it has been oversold, see Jonathan Wells's *Zombie Science: More Icons of Evolution* (Seattle: Discovery Institute Press, 2017), 50–54.

50. Percy B. Shelley, preface to *Frankenstein: Or the Modern Prometheus* [1818], by Mary Shelley, ed. Maurice Hindle (London: Penguin, 2003), 11.

51. Maurice Hindle, introduction to *Frankenstein: Or the Modern Prometheus* [1818], by Mary Shelley, ed. Maurice Hindle (London: Penguin, 2003).

52. Browne, *Charles Darwin: Voyaging*, 39.

53. Francis Crick, *Life Itself: Its Origin and Nature* (New York: Simon and Schuster, 1981). See Michael Denton, *Nature's Destiny: How the Laws of Biology Reveal Purpose in the Universe* (New York: Free Press, 1998), 293.

54. See, for instance, Paul Davies, *The Eerie Silence: Searching for Ourselves in the Universe* (London: Penguin, 2011), 185–195.

55. Dawkins, *The Blind Watchmaker*, 148.

56. Thomas Nagel, *Mind and Cosmos: Why the Materialist Neo-Darwinian Conception of Nature Is Almost Certainly False* (Oxford: Oxford University Press, 2012), 49.

57. James Tour, "We're Still Clueless about the Origin of Life," in *The Mystery of Life's Origin: The Continuing Controversy*, ed. Charles B. Thaxton et al., New Expanded Edition (Seattle: Discovery Institute, 2020), 323–357.

58. See Daniel Dennett, *Darwin's Dangerous Idea: Evolution and the Meanings of Life* (London: Allen Lane, 1995), 314.

59. Michael Denton, *Evolution: A Theory in Crisis* (Bethesda, MD: Adler and Adler, 1986), 271.

60. Fred Hoyle, *The Intelligent Universe* (London: Michael Joseph, 1983), 23.

61. Hoyle, *The Intelligent Universe*, 12.

3. The Challenge of Intelligent Design

1. Paul Davies, *Superforce: The Search for a Grand Unified Theory of Nature* (New York: Simon and Schuster, 1984), 235–6.

2. The term apparently was first used for this purpose by John Herschel in a February 20, 1836, letter to Charles Lyell, and Lyell used it routinely thereafter.

3. Søren Løvtrup, *Darwinism: The Refutation of a Myth* (London: Croom Helm, 1987), 422.

4. David Knight, *Science and Spirituality: The Volatile Connection* (London: Routledge, 2004), 142.

5. Thomas Henry Huxley, "The Origin of Species," *Westminster Review*, April 1860, https://www.gutenberg.org/files/2929/2929-h/2929-h.htm.

6. Quoted in Janet Browne, *Charles Darwin: The Power of Place* (New York: Alfred A. Knopf, 2002), 75–76, 112–113. See also *The Correspondence of Charles Darwin: Volume 7, 1858–1859*, eds. Frederick H. Burkhardt and Sydney Smith (Cambridge: Cambridge University Press, 1991), 289–290.

7. Browne, *Charles Darwin: The Power of Place*, 90–120. For further discussion of the reception of *Origin*, see also A. N. Wilson, *Charles Darwin: Victorian Mythmaker* (London: John Murray, 2017), 245–246, 278–279; and David L. Hull, *Darwin and His Critics: The Reception of Darwin's Theory of Evolution by the Scientific Community* (Chicago: Chicago University Press, 1973).

8. Darwin wrote, "I have heard by round about channel that Herschel says my Book 'is the law of higgledy-piggledy' ... What this exactly means I do not know, but it is evidently very contemptuous—If true this is great blow & discouragement." Charles Darwin to Charles Lyell, December 10, 1859, Darwin Correspondence Project, Letter no. 2575, University of Cambridge, https://www.darwinproject.ac.uk/letter/DCP-LETT-2575.xml.

9. Richard Owen, *The Archetype and Homologies of the Vertebrate Skeleton* (London: Richard and John E. Taylor, 1848).

10. Moritz Wagner, *Die Darwinistische Theorie und das Migrations-Gesetz der Organismen* (Leipzig: Truebner, 1868).

11. St. George Jackson Mivart, *On the Genesis of Species* (New York: Appleton, 1871), 35–75 (chapter on Incipient Structures). The question revolved around what use partially formed limbs or organs could be.

12. Mivart, *On the Genesis of Species*, 34.

13. Edward Hitchcock, *The Religion of Geology* (London: William Collins, 1851).

14. Hull, *Darwin and His Critics*, 302–350.

15. William Bateson, *Problems of Genetics* [1913] (London: Yale University Press, 1979), 248. See also Bateson's *Materials for the Study of Variation with Especial Regard to Discontinuity in the Origin of Species* (London: MacMillan, 1894).

16. Vernon L. Kellogg, *Darwinism To-Day* (New York: Henry Holt, 1907).

17. According to historian Michael Flannery, the notion of a "Darwinian eclipse" and the idea that the neo-Darwinian synthesis "rescued" it from oblivion is a concoction of Julian Huxley and Ernst Mayr, with little real historical evidence to recommend it. For details see Michael A. Flannery, "Toward a New Evolutionary Synthesis," *Theoretical Biology Forum* 110, nos. 1/2 (2017): 47–62, https://pubmed.ncbi.nlm.nih.gov/29687831/. Much of this follows the lead of Ron Amundson, *The Changing Role of the Embryo in Evolutionary Thought: Roots of Evo-Devo* (Cambridge, UK: Cambridge University Press, 2005).

18. Henri Bergson, *Creative Evolution*, trans. Arthur Mitchell (New York: Digireads Publishing, 2011).

19. Pierre-Paul Grassé, *The Evolution of Living Organisms: Evidence for a New Theory of Transformation* (London: Academic Press, 1977), 6, translated from the original, *L'Evolution du Vivant* (Paris: Editions Albin Michel, 1973).

20. Grassé, *The Evolution of Living Organisms*, 246.

21. Medawar developed a number of critiques of scientific overreach, which came together in his *The Limits of Science* (Oxford: Oxford University Press, 1985).

22. Arthur Koestler, *The Ghost in the Machine* (London: Hutchinson, 1967), 129.

23. Raymond Tallis, *Aping Mankind: Neuromania, Darwinitis and the Misrepresentation of Humanity* (Durham: Acumen, 2011), 8.

24. David Holbrook, *Evolution and the Humanities* (Aldershot: Gower, 1987), 192–200.

25. Kim Sterelny, *Dawkins vs. Gould: Survival of the Fittest* (Cambridge: Icon, 2007), 14.

26. Alvar Ellegård, *Darwin and the General Reader: The Reception of Darwin's Theory of Evolution in the British Periodical Press* 1859–72 (Gothenburg: Elanders, 1958).

27. R. L. Trask, *Language*, 2nd ed. (London: Routledge, 1995), 18–19 and figures 1.3 and 1.4 for the physiological structures.

28. Chomsky's idea of an innate universal grammar is not without its prominent detractors. Perhaps the best known is Daniel Everett, who argues that the grammatical feature known as recursion is not in fact universal, contra Chomsky. See Daniel L. Everett, *Language: The Cultural Tool* (New York: Pantheon Books, 2012). For a popular-level overview of this controversy, see Tom Wolfe, *The Kingdom of Speech* (New York: Little Brown and Company, 2016) and Tom Bartlett, "Angry Words," *The Chronicle of Higher Education*, March 20, 2012, https://www.chronicle.com/article/angry-words/.

29. Noam Chomsky, *Language and Mind* (New York: Harcourt Brace Jovanovich, 1972), 97.

30. Alfred Russel Wallace, *The Supernatural Philosopher: Alfred Russel Wallace on Miracles and Skepticism*, eds. Andrea Diem and David C. Lane (Walnut, CA: MSAC Philosophy Group, 2018), 5.

31. "Wallace on Natural Selection," *Nature* (October 13, 1870): 471–473.

32. Charles Darwin to Alfred Russel Wallace, January 26, [1870], in *Alfred Russel Wallace: Letters and Reminiscences*, ed. James Marchant (London: Cassell and Company, 1916), 1:251.

33. For a nuanced account of Alfred Russel Wallace's investigation of seances and spiritualism, see Efram Sera-Shriar, "Credible Witnessing: A. R. Wallace, Spiritualism, and a 'New Branch of Anthropology,'" *Modern Intellectual History* 17, no. 2 (2020): 357–384, doi:10.1017/ S1479244318000331. Two other sources on the topic are Peter Raby, *Alfred Russel Wallace: A Life* (London: Pimlico, 2002), 184–199; and Charles H. Smith, "Wallace, Spiritualism and Beyond: Change or No Change?' in *Natural Selection and Beyond: The Intellectual Legacy of Alfred Russel Wallace*, eds. Charles H. Smith and George Beccaloni (Oxford: Oxford University Press, 2010), 391–423.

34. For how two historians of Darwin apparently got this timeline flipped, see Michael Flannery, "A. N. Wilson's Charles Darwin: The Good, the Bad, and the Ugly," Discovery Institute, December 20, 2017, https://www.discovery.org/a/25402/.

35. Michael Denton, *Nature's Destiny: How the Laws of Biology Reveal Purpose in the Universe* (New York: Free Press, 1998), 348.

36. Susan Blackmore, *Consciousness: An Introduction* (London: Hodder and Stoughton, 2003).

37. Michael Denton, *Evolution: A Theory in Crisis* (Bethesda, MD: Adler and Adler, 1986), 358.

38. From the mid-1860s onwards Wallace argued for a special place for the human mind, concluding that a higher power must have been behind its origin (rather than natural selection). The human mind was the living proof of a divine mind, on the Platonic principle that—excepting the First Cause—everything requires a blueprint or original exemplar. This First Cause was instantiated through teleological laws of nature, Wallace's burgeoning religious views being an inference from an observation of nature itself. According to his position, that divine power cannot be directly apprehended but can still be inferred from the powers of nature. Hence the awe we feel in the presence of nature which, in the thesis of early twentieth-century German theologian Rudolf Otto, is essentially in and of itself an apprehension of the Holy. Wallace's evolving conception of what could or could not be expected of natural selection makes him something of a prophet for—or at least a forerunner of—the modern intelligent design movement in scientific thought, which also denies that irreducibly complex biological systems could have been thrown together by the undiscriminating push and pull of blind forces, with or without natural selection. For more on Wallace's development of a teleological model of evolution, see Neil Thomas, *The God Paradigm Revisited. Darwin and Wallace: Natural Selection and Natural Philosophy* (London: Amazon Publishing, 2021), as well as Alfred Russel Wallace, *The World of Life: A Manifestation of Creative Power, Directive Mind and Ultimate Purpose* (London: Chapman & Hall, 1914); and Michael Flannery, *Intelligent Evolution: How Alfred Russel Wallace's "World of Life" Challenged Darwinism* (Riesel, TX: Erasmus Press, 2020). Rudolf Otto's *Das Heilige* first appeared in German in 1917, and six years later in English as *The Idea of the Holy* [1923] (Oxford: Oxford University Press, 1958).

39. Fred Hoyle, *The Intelligent Universe: A New View of Creation and Evolution* (London: Michael Joseph, 1983), 1.

40. Lee Spetner, *Not by Chance! Shattering the Modern Theory of Evolution* (New York: Judaica Press, 2006), 166.

41. John Reader, *Missing Links: The Hunt for Earliest Man* (London: Penguin, 1988), xiv.

42. M. Bowden, *Ape-Men, Fact or Fallacy? A Critical Examination of the Evidence* (Bromley, Kent: Sovereign Publications, 1977). For an unbiased account of this and other forgeries see Reader, *Missing Links*, especially 54–78 (on Piltdown Man).

43. Gertrude Himmelfarb, *Darwin and the Darwinian Revolution* (London: Chatto and Windus, 1959), 310–11.

44. Charles Darwin, *On the Origin of Species* [1859], ed. Gillian Beer (Oxford: Oxford University Press, 2008), 129.

45. Himmelfarb, *Darwin and the Darwinian Revolution*, 278.

46. Cited by Thomas Woodward, *Doubts about Darwin, A History of Intelligent Design* (Grand Rapids: Baker Books, 2003), 43.

47. Charles Darwin, *The Descent of Man, and Selection in Relation to Sex* [1871], eds. James Moore and Adrian Desmond (London: Penguin, 2004).

48. Darwin, *The Descent of Man*, 107.

49. Darwin, *The Descent of Man*, 119.

50. Himmelfarb, *Darwin and the Darwinian Revolution*, 307.

51. James Moore and Adrian Desmond, introduction to *The Descent of Man, and Selection in Relation to Sex* [1871], by Charles Darwin, eds. James Moore and Adrian Desmond (London: Penguin, 2004), xxvi.

52. Keith Thomas, *Man and the Natural World: Changing Attitudes in England 1500–1800* (London: Penguin, 1983), 141.

53. Charles Darwin, "Notebook C [1838]," entry 79, in *Charles Darwin's Notebooks 1836–1844*, ed. Paul H. Barrett et al. (New York: Cambridge University Press, 1987), 264.

54. Himmelfarb, *Darwin and the Darwinian Revolution*, 307–8.

55. The phrase is that of Roger Kimball in his introduction to David Stove's *Darwinian Fairy Tales: Selfish Genes, Errors of Heredity and Other Fables of Evolution* (New York: Encounter, 1995), xiii.

56. Antony Flew, *Darwinian Evolution*, 2nd ed. (London: Transaction, 1997), 25. See also Mary Midgley, *Evolution as Religion* (London: Routledge, 2002).

57. Himmelfarb, *Darwin and the Darwinian Revolution*, 308.

58. On Kropotkin see John Hands, *Cosmosapiens: Human Evolution from the Origin of the Universe* (London: Duckworth, 2015), 267–70.

59. See Penny Spikins, *How Compassion Made Us Human: The Evolutionary Origins of Tenderness, Trust and Morality* (Barnsley: Pen and Sword, 2015).

60. Darwin, *The Descent of Man*, 110.

61. Darwin, *On the Origin of Species*, 337.

62. Bateson, *Materials for the Study of Variation*, 5.

63. A squib aimed at the consummate watchmaker-Creator of Paley's *Natural Theology*.

64. Dawkins, *The Blind Watchmaker*, 162.

65. Neil Broom, *How Blind is the Watchmaker? Nature's Design and the Limits of Naturalistic Science* (Downers Grove, IL: Intervarsity Press, 2001), 73. In commenting on the sleights of hand among evolutionists, Broom further notes, "Explanations in biology are often couched in terms of things having 'evolved' from simpler systems, thereby giving the impression that no deeper explanation is required. But this kind of talk has no more explanatory power than is contained in the statement 'the modern computer evolved from the Chinese abacus.' It might be historically true but it tells us absolutely nothing about the crucial role of human creativity and ingenuity in this technological evolution." See 39, note 12.

66. Thomas Nagel, *Mind and Cosmos: Why the Materialist Neo-Darwinian Conception of Nature Is Almost Certainly False* (Oxford: Oxford University Press, 2012), 12.

4. Cosmos and Chaos

1. Charles H. Townes, "Explore As Much As We Can," interview by Bonnie Azab Powell, UCBerkeleyNews, June 17, 2005, https://www.berkeley.edu/news/media/releases/2005/06/17_townes.shtml.

2. Bertrand Russell, *Why I Am Not a Christian* [1927] (New York: Simon and Schuster, 1957), 107.

3. Richard Dawkins, *River Out of Eden: A Darwinian View of Life* (New York: Basic Books, 1995), 133.

4. This was pointed out as early as the 1970s by philosopher Richard Spilsbury in *Providence Lost: A Critique of Darwinism* (Oxford: Oxford University Press, 1974), especially 111–130.

5. Bertrand Russell, *Religion and Science* (London: Thornton Butterworth, 1935), 221–222.

6. Brandon Carter, "Large Number Coincidences and the Anthropic Principle in Cosmology," Symposium—International Astronomical Union 63 (1974): 291–98, doi:10.1017/S0074180900235638.

7. Darwin's tutor, William Whewell, had made reference to the uncanny suitability of the planet for life and so had Alfred Russel Wallace in his *Man's Place in the Universe*, published in 1903. See Michael A. Flannery, *Alfred Russel Wallace: A Rediscovered Life* (Seattle: Discovery Institute Press, 2011), 87–89. See also a groundbreaking work on the subject by Lawrence J. Henderson, *The Fitness of the Environment* (New York: MacMillan, 1913).

8. See Guillermo Gonzalez and Jay Richards, *The Privileged Planet* (Washington, DC: Regnery, 2004); Peter Ward and Donald Brownlee, *Rare Earth* (New York: Copernicus, 2000); and David Waltham, *Lucky Planet* (New York: Basic Books, 2014).

9. Brandon Carter, "Large Number Coincidences," 291, 293.

10. Paul Davies develops this point further in several books, including *The Accidental Universe* (Cambridge, UK: Cambridge University Press, 1982), *God and the New Physics* (London: Penguin, 1990); *The Fifth Miracle: The Search for the Origin of Life* (London: Penguin, 1999); *The Goldilocks Enigma: Why Is the Universe Just Right for Life?* (London: Penguin, 2007); and *The Eerie Silence: Searching for Ourselves in the Universe* (London: Penguin, 2010).

11. George Gaylord Simpson, *The Meaning of Evolution: Revised Edition* (New Haven: Yale University Press, 1967), 345.

12. Some historians question whether the shift from a geocentric cosmology to a heliocentric one actually demoted the Earth in the eyes of Renaissance scholars. In the Medieval geocentric cosmology that Copernicus challenged, Earth was regarded as the bottom, or sump, of the universe. On this view, the heliocentric model could be said to have been a promotion for planet Earth. For more analysis of this issue, see Michael Newton Keas, *Unbelievable: 7 Myths About the History and Future of Science* (Wilmington, DE: ISI Books, 2019), 91–107.

13. Michael Denton, *Nature's Destiny: How the Laws of Biology Reveal Purpose in the Universe* (New York: Free Press, 1998), 370.

14. See Stephen J. Gould, *Rock of Ages: Science and Religion in the Fullness of Life* (New York: Ballantine, 1999).

15. Jacques Barzun, *Darwin, Marx, Freud: Critique of a Heritage*, rev. 2nd ed. (New York: Doubleday Anchor, 1958), 51.

16. William Whewell, *Of the Plurality of Worlds* (London: John Parker and Son, 1853), 206.

17. David Waltham, *Lucky Planet: Why Earth is Exceptional—And What That Means for Life in the Universe* (New York: Basic Books, 2014), 13.

18. In his seminal 1973 paper, Carter himself suggested an ensemble of universes as a way of explaining examples of fine tuning in our universe, though he conceded that such an explanation

was not his favored choice, just one that ought to be considered. See Carter, "Large Number Coincidences," 298.

19. Rupert Sheldrake, *The Science Delusion* (London: Hodder and Stoughton, 2013), 12.

20. Stephen C. Meyer, *Return of the God Hypothesis: Three Scientific Discoveries that Reveal the Mind Behind the Universe* (San Francisco: HarperOne, 2021), 339–343. After reviewing several ways that proposed multiverse-generating mechanisms themselves require exquisite fine tuning, Meyer concludes thus: "Robin Collins has a clever way of characterizing this whole situation. He likens physicists who attempt to explain fine tuning solely by reference to universe-creating mechanisms, without intelligent design, to a hapless soul who denies any human ingenuity in the making of a freshly baked loaf of bread simply because the baker used a breadmaking machine. Clearly, argues Collins, such a benighted fellow has overlooked an obvious fact: the breadmaking machine itself required prior ingenuity and design, as did the recipe for and the preparation of the dough that went into it. Similarly, even if a multiverse hypothesis is true, it would support, rather than undermine, the intelligent design hypothesis, since the multiverse hypothesis depends upon the specific features of universe-generating mechanisms that invariably require prior and otherwise unexplained fine tuning." [internal reference removed]

21. St. George Jackson Mivart, *On the Genesis of Species* (New York: Appleton, 1871), 12.

22. August Weismann, "The All-Sufficiency of Natural Selection," *Contemporary Review 1866–1900*, 64 (September 1893): 309–338, https://archive.org/details/1893-weismann. The portions quoted here appear on page 328 and 336.

23. Richard C. Lewontin, "Billions and Billions of Demons," *New York Review*, January 9, 1997, 31, https://www.nybooks.com/articles/1997/01/09/billions-and-billions-of-demons/.

24. Charles Coulston Gillispie, *Genesis and Geology: A Study in the Relations of Scientific Thought, Natural Theology, and Social Opinion in Great Britain, 1790–1850* (Chicago: Chicago University Press, 1979), 30.

25. Paul Davies, *The Mind of God: The Scientific Basis for a Rational World* (New York: Simon & Schuster, 1992), 16.

26. Antony Flew, "My Pilgrimage from Atheism to Theism: A Discussion between Antony Flew and Gary Habermas," *Philosophia Christi* 6, no. 2 (2004): 197–211, http://epsociety.org/library/articles.asp?pid=33&ap=1.

27. William Paley, *Natural Theology* (New York: American Tract Society, n.d., ca. 1805), 20.

28. Douglas Dewar, *More Difficulties of the Evolution Theory* (London: Thynne and Co., 1938), 5.

29. Antony Flew, *There Is a God* (New York: Harper Collins, 2007), 85.

30. Readers interested in this subject can find a good account in Edward J. Larson's *Summer for the Gods: The Scopes Trial and America's Continuing Debate over Science and Religion* (New York: Basic Books, 1997). For an article available online that highlights the heavy bias and inaccuracies of the film, see Carol Iannone, "The Truth about *Inherit the Wind*," *First Things*, February 1997, https://www.firstthings.com/article/1997/02/002-the-truth-about-inherit-the-wind--36.

5. THE MYSTERY OF MYSTERIES

1. Fern Elsdon-Baker, *The Selfish Genius: How Richard Dawkins Rewrote Darwin's Legacy* (London: Icon, 2009), 14.

2. Noam Chomsky, *Reflections on Language* (New York: Pantheon, 1975), 157. On this subject see also the overview in Daniel Dennett, *Darwin's Dangerous Idea: Evolution and the Meanings of Life* (London: Allen Lane, 1995), 381–400.

3. Charles Darwin to George Charles Wallace, March 28, 1882, in Gavin De Beer, "Some Un-published Letters of Charles Darwin," *Notes and Records of the Royal Society of London* 14, no. 1 (June 1959): 59.

4. John C. Lennox, *God and Stephen Hawking: Whose Design Is It Anyway?* (Oxford: Lion Hudson, 2010), 41–42.

5. Charles Darwin to George Charles Wallace, March 28, 1882, in Gavin De Beer, "Some Un-published Letters of Charles Darwin," *Notes and Records of the Royal Society of London* 14, no. 1 (June 1959): 59.

6. Charles Darwin, *On the Origin of Species* (London: John Murray, 1859), 488, Darwin Online, http://darwin-online.org.uk/content/frameset?pageseq=506&itemID=F373&viewtype=image.

7. Darwin, *On the Origin of Species*, 490.

8. Neal C. Gillespie, *Charles Darwin and the Problem of Creation* (Chicago: Chicago University Press, 1979), 125.

9. Jacques Barzun, *Darwin, Marx, Wagner: Critique of a Heritage*, 2nd ed. (New York: Doubleday Anchor, 1958), 76–77.

10. Charles Darwin to Charles Lyell, October 11, 1859, Darwin Correspondence Project, Letter no. 2503, University of Cambridge, https://www.darwinproject.ac.uk/letter/DCP-LETT-2503.xml.

11. Charles Darwin, *The Autobiography of Charles Darwin 1809-1882*, ed. Nora Barlow [1958] (London: Norton, 1993), 92–93.

12. Vernon L. Kellogg, *Darwinism To-Day* (New York: Henry Holt, 1907).

13. On this subject see Peter Bowler, *The Eclipse of Darwinism: Anti-Darwinian Evolution Theory in the Decades around 1900*, 2nd ed. (Baltimore: Johns Hopkins University Press, 1992).

14. For an overview of the challenges facing these extensions and evolutionary alternatives to neo-Darwinism, see Michael J. Behe, *Darwin Devolves: The New Science about DNA that Challenges Evolution* (San Francisco: HarperOne, 2019), chapters 4 and 5; Stephen C. Meyer, *Darwin's Doubt: The Explosive Origin of Animal Life and the Case for Intelligent Design* (San Francisco: HarperOne, 2013), chapters 15 and 16; and *Theistic Evolution: A Scientific, Philosophical, and Theological Critique* (Wheaton, IL: Crossway, 2017), chapter 8.

15. Rupert Sheldrake, *The Science Delusion* (London: Hodder and Stoughton, 2013), 9. This was a private communication from Crick's son to Sheldrake.

16. Fred Hoyle, *The Intelligent Universe: A New View of Creation and Evolution* (London: Michael Joseph, 1983), 19.

17. Richard Dawkins, *The Blind Watchmaker* [1986] (New York: W. W. Norton, 1996), 226.

18. Peter Atkins, *On Being: A Scientist's Exploration of the Great Questions of Existence* (Oxford, UK: Oxford University Press, 2011), 21.

19. Hodge reportedly made this remark in response to a presentation by the internationally renowned philosopher James McCosh during the Evangelical Alliance conference, New York, October 2–12, 1873. McCosh attempted to reconcile Darwinism and the Bible; in response Hodge summarized the arguments he would publish the following year in *What Is Darwinism?* (New York: Scribner, 1874). For an account of the convention debate see George Marsden, *Fundamentalism and American Culture* (New York: Oxford University Press, 1980), 17–19; see also 232, note 3.

20. Albert Einstein to Phyllis Wright, January 24, 1936. Translated from the original German in Max Jammer, *Einstein and Religion: Physics and Theology* (Princeton, NJ: Princeton University Press, 1999), 92–93.

21. James Le Fanu, *Why Us?* (London: Harper, 2009), 58.

22. Linda Gamlin, "The Human Immune System: Origins," *The New Scientist inside Science*, ed. Richard Fifield (London: Penguin, 1992), 225–226.

23. As an example of the poetic technique known as the pathetic fallacy, in Act 2, scene 3 of *Macbeth*, the character of Lennox, a Scottish nobleman, says: "The night has been unruly. Where we lay, / Our chimneys were blown down and, as they say, / Lamentings heard i' th' air, strange screams of death, / And prophesying with accents terrible / Of dire combustion and confused events / New hatched to the woeful time. The obscure bird / Clamored the livelong night. Some say the Earth / Was feverous and did shake."

24. The seventeenth-century philosopher Spinoza referred to this active conception of Nature as an active *natura naturans*, whereas the more modern conception of Nature as inert and passive he termed *natura naturata*.

25. For Wallace at least it would be going too far to say he was long intent on an anti-theistic model. In his 1856 article titled "On the Habits of the Orang-Utan of Borneo" in *Annals and Magazine of Natural History* 2nd ser., vol. 17, no. 103, he is already hinting at the theistic direction he was contemplating and would eventually take. Historian Michael Flannery writes, "It is no exaggeration to see this 1856 essay, written in the wake of his Sarawak Law paper the year before and ahead of his famous Ternate letter, as an early creedal statement. It would mark the emergent tenets of his inchoate teleological worldview, which consisted of the following: a nonreductionist, holistic view of nature; an admission of inutility in the plant and animal kingdoms and this given as reasonable evidence of higher and even intelligent causation in nature; a special place for humankind in the appreciation of features beyond mere survival utility such as beauty of form, color, and majesty; and the allowance that all of this may be the intentional expression of a theistic presence or force." Flannery, *Nature's Prophet: Alfred Russel Wallace and His Evolution from Natural Selection to Natural Theology* (Tuscaloosa, AL: University of Alabama Press, 2018), 64.

26. Samuel Wilberforce, "On the Origin of Species," *Quarterly Review* (1860), 237–238.

27. Regarding natural selection, historian Michael Flannery notes (in private correspondence) that a major point of difference that came to separate Wallace from Darwin was the question of artificial selection's evidential import. In *On the Origin of Species* Darwin offered breeding examples as analogous to natural selection. Wallace, from his Ternate paper (1858) on, clearly distinguished between the two, and by implication limited the explanatory power of natural selection. We find this in his paper presented to the Anthropological Society of London in 1864, and it led to his open break with Darwin in 1869.

28. Darwin, *On the Origin of Species*, 141–142.

29. Antony Flew, *Darwinian Evolution*, 2nd ed. (London: Transaction Publishers, 1997), 25. Bombay duck is a gastronomic delicacy composed of dry, salted fish.

30. Le Fanu, *Why Us?*, 107.

31. On this point see now John Hands, *Cosmosapiens: Human Evolution from the Origin of the Universe* (London: Duckworth, 2015), 382.

32. See Nancy Pearcey, "You Guys Lost," in *Mere Creation: Science, Faith and Intelligent Design*, ed. William A. Dembski (Downers Grove, IL: Intervarsity Press, 1998), 73–92.

33. Charles Lyell, *Sir Charles Lyell's Scientific Journals on the Species Question*, ed. Leonard G. Wilson (New Haven, CT: Yale University Press, 1970), 56–57, 84.

34. John William Draper, *History of the Conflict between Religion and Science* (New York: Appleton, 1878).

35. Auguste Comte, *Cours de Philosophie Positive* [1830 and 1835]. Reprinted in English in Comte's *Philosophy of the Sciences: Being an Exposition of the "Cours de Philosophie Positive" of*

Auguste Comte, ed. G. H. Lewis (London: G. Bohn, 1853), 88, https://archive.org/details/dli. granth.41377/page/88/mode/2up.

6. Paradigms Regained

1. Søren Løvtrup, *Darwinism: The Refutation of a Myth* (London: Croom Helm, 1987), 422.

2. Allan Sandage, quoted by John Noble Wilford, "Sizing Up the Cosmos: An Astronomer's Quest," March 12, 1991, *New York Times*, https://www.nytimes.com/1991/03/12/science/ sizing-up-the-cosmos-an-astronomer-s-quest.html.

3. Harold Jeffreys, *The Earth: Its Origin, History and Physical Constitution* (Cambridge: Cambridge University Press, 2008), 359.

4. James Le Fanu, *Why Us?* (London: Harper, 2009), 122.

5. Paul Davies, *The Mind of God: Science and the Search for Ultimate Meaning* (London: Penguin, 1992), 15–16.

6. Peter Bowler, *Darwin Deleted: Imagining a World Without Darwin* (Chicago: Chicago University Press, 2013), 276.

7. Neo-Darwinism emerged as the "winner" after Mendelian genetics disproved Lamarck's theory that acquired characteristics could be passed down to future generations.

8. Stephen Jay Gould, "Evolution's Erratic Pace," *Natural History*, 86, no. 5 (May 1977), 14.

9. Murray Eden, "Inadequacies of Neo-Darwinian Evolution as a Scientific Theory," in *Mathematical Challenges to the Neo-Darwinian Interpretation of Evolution*, eds. Paul S. Moorhead and Martin M. Kaplan (Philadelphia: Wistar Institute Press, 1967), 109.

10. Norman Macbeth, *Darwin Retried* (London: Garnstone, 1971), 6–7.

11. Thomas Dixon, *Science and Religion: A Very Short Introduction* (Oxford: Oxford University Press, 2008), 98.

12. Scientists' pulling of rank and obstructiveness towards non-standard thinking is not a point I wish to foreground here since I deem distracting *personalia* irrelevant to the main issue, but interested persons may consult Richard Milton, *Shattering the Myths of Darwinism* (London: Fourth Estate, 1997), especially 265–272, and Matti Leisola and Jonathan Witt, *Heretic: One Scientist's Journey from Darwin to Design* (Seattle: Discovery Institute Press, 2018).

13. For an overview of these developments see Oliver Robinson, *Paths between Head and Heart: Exploring the Harmonies of Science and Spirituality* (Washington, DC: O-Books, 2018), 157–61.

14. Peter Bowler, *The Eclipse of Darwinism* (Baltimore: Johns Hopkins University Press, 1983), 57.

15. Michael A. Flannery, *Alfred Russell Wallace: A Rediscovered Life* (Seattle: Discovery Institute Press, 2011), 110.

16. For an accessible sampler on this subject see Theodore Dalrymple and Kenneth Francis, *The Terror of Existence from Ecclesiastes to the Theatre of the Absurd* (London: New English Review Press, 2018).

17. Richard Swinburne, quoted in Andrew Zak Williams, "I'm a Believer," *New Statesman*, April 20, 2011, https://www.newstatesman.com/religion/2011/04/god-believe-faith-world-belief. Williams wrote to dozens of scientists and other public figures, requesting explanations for their faith. Swinburne was one of the correspondents who responded.

18. Philip G. Fothergill, *Historical Aspects of Organic Evolution* (London: Hollis and Carter, 1952), 344.

19. Davies, *The Mind of God*, 15.

20. Peter J. Vorzimmer, *Charles Darwin: The Years of Controversy: The Origin of Species and its Critics 1859–1882* (London: University of London University Press, 1972), 224.

21. Charles Darwin to Asa Gray, May 22, 1860, Darwin Correspondence Project, Letter no. 2814, University of Cambridge, https://www.darwinproject.ac.uk/letter/DCP-LETT-2814.xml.

22. Charles Darwin, *The Autobiography of Charles Darwin 1809-1882*, ed. Nora Barlow [1958] (London: Norton, 1993), 90–94.

23. Charles Darwin to John Fordyce, May 7, 1879, Darwin Correspondence Project, Letter no. 12040, University of Cambridge, https://www.darwinproject.ac.uk/letter/DCP-LETT-12040.xml.

24. Cited by Maurice Hindle, introduction to *Frankenstein: Or the Modern Prometheus*, by Mary Shelley, ed. Maurice Hindle, 2nd ed. (London: Penguin, 2003), xxix–xxx.

25. David Cromwell et al., *This is Planet Earth: Your Guide to the World We Live In* (London: John Murray, 2018), 175.

26. Neal C. Gillespie, *Charles Darwin and the Problem of Creation* (Chicago: Chicago University Press, 1979), 30.

27. Harold C. Urey, interview with *Christian Science Monitor*, January 4, 1962, 4. Urey, having conceded that life appears "too complex to have evolved anywhere," further confesses that he and his colleagues persist in believing that "life evolved from dead matter on this planet" purely "as an article of faith."

28. Robert Shapiro, *Origins: A Skeptic's Guide to the Creation of Life on Earth* (New York: Bantam, 1987), 13. Shapiro is known for his "metabolism first" model of the spontaneous origin of life, in opposition to the "RNA first" model, which he may have had in mind when he said "some theories come labeled as The Answer."

29. Richard Spilsbury, *Providence Lost: A Critique of Darwinism* (Oxford: Oxford University Press, 1974), 19.

EPILOGUE

1. See Jonathan Wells, *Icons of Evolution: Science or Myth? Why Much of What We Teach about Evolution is Wrong* (Washington, DC: Regnery, 2000). For example, in Chapter 5 Wells highlights how nineteenth-century German biologist Ernst Haeckel's embryo drawings frequently appear in high school biology textbooks as evidence for evolution, despite having been long ago discredited. As Wells notes, they are highly imaginative and inaccurate renderings—retouched, if you will, so as to go well beyond the evidence Haeckel aspired to adduce.

2. "Richards Dawkins on Religion, American Education and Islam," interview by Mark Urban, BBC Newsnight, September 19, 2019, video, 8:46, https://www.bbc.co.uk/programmes/p07nq3ll.

INDEX

Printed in Great Britain
by Amazon